T0343987

Semiconductor Glossary

A Resource for Semiconductor Community

Second Edition

Semiconductor Glossary

A Resource for Semiconductor Community

Second Edition

Jerzy Ruzyllo

Penn State University, USA

 **World Scientific**

NEW JERSEY · LONDON · SINGAPORE · BEIJING · SHANGHAI · HONG KONG · TAIPEI · CHENNAI · TOKYO

Published by

World Scientific Publishing Co. Pte. Ltd.

5 Toh Tuck Link, Singapore 596224

USA office: 27 Warren Street, Suite 401-402, Hackensack, NJ 07601

UK office: 57 Shelton Street, Covent Garden, London WC2H 9HE

British Library Cataloguing-in-Publication Data
A catalogue record for this book is available from the British Library.

Information in this glossary, either numerical or conceptual, is provided at the author's discretion. Author disclaims any liability based on or related to the contents of this glossary.

SEMICONDUCTOR GLOSSARY — 2nd Edition
A Resource for Semiconductor Community

To order additional copies or to inquire about discounts on bulk quantities contact us at:
www.worldscientific.com

ISBN 978-981-4749-53-4
ISBN 978-981-4749-54-1 (pbk)

Printed in Singapore

Over the last six decades semiconductor science and engineering has grown into one of the cornerstones of our technical civilization. It continues to grow and expand into new territories at an accelerating pace, which makes keeping up with emerging concepts and terminology increasingly challenging. *Semiconductor Glossary* was conceived in response to an apparent need for reference that would provide brief, straightforward explanations of key terms in the area of semiconductor materials, devices, and processing.

In agreement with its inherent nature, this glossary is not intended to provide in-depth explanations of complex technical ideas and scientific concepts. Readers searching for this type of knowledge should refer to textbooks and monographs covering all aspects of broadly understood semiconductor science and engineering. The purpose of this glossary is to provide a platform upon which basic concepts and terms in semiconductor science and engineering are identified and explained in an easy to follow fashion.

This volume is a vastly expanded and updated second edition of *Semiconductor Glossary*. An abbreviated version of this material is available on line at *www.semiconductorglossary.com*. The information presented in this volume is based primarily on the knowledge and experiences acquired by the author during over 40 years of research and teaching in the area of semiconductors.

The author would like to acknowledge collaborators and students at Penn State University and the Warsaw University of Technology, as well as numerous colleagues in the industry and academia with whom he has had the privilege of interacting throughout the years. They are the main reason for which immersion in the "world of semiconductors" continues to be such a stimulating and gratifying experience.

Jerzy Ruzyllo
University Park, Pennsylvania

A

a   see *lattice constant.*

a-Si   see *amorphous silicon.*

AII-BVI, II-VI, semiconductors   the II-VI compound semiconductors are synthesized using elements from group II and group VI of the periodic table; e.g., CdSe, ZnS.
elemental semiconductor, CdSe, CdTe, CdHgTe, ZnS, ZnO

AIII-BV, III-V, semiconductors   the III-V compound semiconductors are synthesized using elements from group III and group V of the periodic table; e.g. GaAs, InP.
elemental semiconductor, GaAs, GaN, GaP, GaAlAs, InAs, InSb, InP

AIV-BIV, IV-IV semiconductors   compound semiconductors with both elements forming a compound originating from group IV of the periodic table; e.g. SiC, SiGe.
elemental semiconductor, SiC, SiGe

AAS Atomic Adsorption Spectroscopy; a surface characterization method.

abrupt heterojunction   a junction between two crystalline semiconductors featuring different electron affinities and energy bandgaps; no graded changes in the materials composition in the junction region.
electron affinity, graded heterojunction

abrupt junction   a *p-n* junction in which dopant concentration changes from *p*-type to *n*-type over the very short distance, i.e. transition region between *p*-type and *n*-type parts of the junction is very narrow.
linear junction

absorption   specie or energy penetrates the surface and is bound or captured in the bulk of the solid where it is releasing its energy.
phonon, photon

absorption coefficient, *α*, [cm^{-1}] defines depth of penetration of a given medium with the light of a given wavelength; increases as wavelength is getting shorter.
radiation wavelength-energy conversion

acceptor *p*-type dopant; element introduced to semiconductor to generate free holes (by "accepting" electron from semiconductor atom and "releasing" hole at the same time); acceptor atom must have one valence electron less than the host semiconductor; boron (B) is commonly used acceptor in silicon technology; alternatives include indium (In) and gallium (Ga); gallium features high diffusivity in SiO_2, and hence, oxide cannot be used as the mask during Ga diffusion.
donor, doping

access time time needed for the bit of information to go to and return from the memory cell.
memory cell

accumulation condition of semiconductor surface region in MOS devices in which concentration of majority carriers is higher than concentration of dopant atoms.
depletion, inversion

activation energy defines reaction kinetics of the process; the minimum amount of energy required to initiate the reaction; expressed in units of *eV*.

active element (component) a device which displays asymmetric current-voltage characteristics, i.e. dependent upon the direction of the applied bias; is introducing net energy into the circuit; diodes and transistors are active elements.
passive element, diode, transistor

active Si layer a single-crystal Si layer overlaying buried oxide (BOX) in SOI wafers; "active" because transistors are built into it; as opposed to Si substrate (Si underneath BOX) which is a part of the SOI wafer providing mechanical support only; can be as thin as < 10 nm.
Silicon-on-Insulator; Ultra-Thin Body SOI, BOX, fully-depleted SOI

ADC Analog to Digital Converter.

additive process a process which adds material to the substrate, e.g. in the form of thin-film.
subtractive process

2

adhesion ability of materials to stick (adhere) to each other.
adhesion promoter

adhesion promoter a compound used to improve adhesion of materials; typically understood as a material improving adhesion of the photoresist to the wafer surface in the lithographic processes.
HMDS

adsorption binding between foreign molecules and the solid occurring only on the solid surface; specie is attached to the solid surface by weak physical forces (Van der Waals force).
desorption, van der Waals force

aerosol cleaning removal of the particulate contaminants from the wafer surface using frozen gas particles.
cleaning, dry cleaning, particles

AFM see *Atomic Force Microscopy.*

afterglow plasma, afterglow plasma generated radiation and ions which remain active downstream from the plasma; a plasma processing mode in semiconductor manufacturing.
downstream plasma, remote plasma

ALD, ALCVD Atomic Layer Deposition; Atomic Layer Chemical Vapor Deposition; see *Atomic Layer Deposition.*

AlGaAs GaAs with Al added in adequate amount to modify width of the energy gap; by gradually (layer-by-layer) varying Al content in GaAlAs continued variation of the energy gap of the material is accomplished.
bandgap engineering, GaAs

aligner a tool used in photolithography which allows desired positioning of the mask (or reticle) relative to a wafer prior to exposure of the photoresist; "aligner" is at the same time an exposure tool.
exposure, photolithography

alignment process of positioning of the mask (or reticle) relative to the wafer prior to exposure of the photoresist in photolithographic processes.
exposure, mask, reticle

alignment mark specially configured mark put on each mask to allow its precise alignment with the pattern on the wafer.
alignment

alloyed junction a junction formed by alloying metal acting as a dopant with semiconductor for the purpose of *p-n* junction formation; e.g. alloy of indium with *n*-type Si forms *p*-type region of the *p-n* junction.
diffused junction, implanted junction

alternative dielectrics dielectrics featuring dielectric constant $k > 3.9$ (3.9 is a dielectric constant of SiO_2) and acting as gate dielectrics in silicon MOS devices instead of SiO_2; referred to as "high-k dielectrics"; also dielectrics featuring dielectric constant $k < 3.9$ and used as ILD; referred to as "low-k dielectrics".
high-k dielectric, low-k dielectric, ILD

aluminum, Al, conductor common metal in semiconductor processing; used for contacts and interconnects; very low resistivity (2.7 $\mu\Omega$-cm); melting point 660 °C; easy deposition by evaporation or sputtering; easy etching; shortcomings: electromigration, spiking of silicon, insufficient temperature resistance.
electromigration, spiking

aluminum, Al, contaminant common metallic contaminant in silicon processing; main source: APM cleaning solution and water; slows down thermal oxidation of silicon; affects oxide reliability; detection and measurement on Si surface e.g. by TOF-SIMS.
APM, TOF-SIMS

aluminum nitride, AlN a wide-bandgap semiconductor ($E_g = 6.2$ eV); features direct energy gap, used in UV detection devices.
boron nitride

aluminum oxide, alumina, Al_2O_3 oxide featuring energy gap $E_g \sim 5$ eV and dielectric constant $k \sim 8$; in the form of a single-crystal known as sapphire.
sapphire

ambipolar diffusion coefficient the effective diffusion coefficient of the excess charge carriers (electrons and holes) in semiconductor.
diffusion coefficient

ambipolar mobility the effective mobility of the excess charge carriers (electronics and holes) in semiconductor.
mobility

ambipolar transport the situation where excess electron and hole concentrations in semiconductor are nearly equal is known as quasi-neutrality condition; the charge transport in this case is described by the so-called ambipolar transport in which electrons and holes are diffusing or drifting with the same ambipolar diffusion coefficient and ambipolar mobility.
excess carriers

AMLCD Active Matrix Liquid Crystal Display; higher performance version of an LCD; uses arrays of thin film transistors (TFT) to control individual pixels.
Thin Film Transistor, Liquid Crystal Display

ammonia, NH_3 a gas used in semiconductor processing as a source of atomic nitrogen; atomic nitrogen can be extracted from ammonia at lower temperature than from the molecular nitrogen, N_2; toxic.
nitridation

ammonium hydroxide, NH_4OH a liquid obtained by dissolving ammonia in water; key ingredient of the APM cleaning solution.
APM

AMOLED Active Matrix Organic Light Emitting Diode, active matrix display constructed using organic LEDs in conjunction with thin-film transistors.
organic semiconductor, LED, AMLCD

amorphous material a non-crystalline solid; features no periodicity and long-range order; lower quality than crystalline materials, but cheaper to form; amorphous semiconductors are useful in large-area applications such as solar cells and flat-panel displays; insulators used in semiconductor technology, e.g. SiO_2 and Si_3N_4, are amorphous.
single-crystal, poly-crystalline material

amorphous Si, a-Si a non-crystalline thin-film silicon; features no long-range crystallographic order; inferior electrical characteristics as

compared to single-crystal and polycrystalline-Si (poly-Si), but cheaper and easier to manufacture; used primarily to fabricate solar cells and thin-film transistors.
hydrogenated a-Si, solar cell, TFT

amphoteric dopant an element which can act either as a donor or an acceptor in a given semiconductor; e.g. Si is an amphoteric dopant in GaAs where it acts as a donor on Ga site in GaAs structure or as an acceptor when positioned on As site.
acceptor, donor, dopant

analog device a device designed to realize analog functions such as signal amplification; in analog devices output signal follows continuously input signal.
digital device

analog integrated circuit an integrated circuit realizing analog functions; e.g. operational amplifier; in analog integrated circuits, in contrast to digital integrated circuits, output signal follows continuously input signal.
digital integrated circuit

angstrom, Å unit of length; often used in semiconductor nomenclature; $1Å = 10^{-8}$ cm $= 10^{-4}$ μm $= 0.1$ nm; not a standard international (SI) unit; replaced by nm in common usage (1 nm $= 10Å$).
nanometer

anhydrous HF, AHF a water-free vapor of hydrofluoric acid; used in conjunction with vapor of either water or alcoholic solvents (e.g. methanol) to etch silicon oxides on the wafer surface or sacrificial oxide in MEMS release processes; oxide etch rate depends strongly on wafer temperature and vapor pressure as well as type of oxide; e.g. dry thermal oxide etches much slower than CVD oxide.
hydrofluoric acid, sacrificial oxide, MEMS release

anisotropic etch etching process in which etch rate in the direction normal to the surface is much higher than in directions parallel to the surface; no undercutting, i.e. lateral distortion of the pattern is minimized; needed to define very tight geometrical features.
isotropic etch, etch rate

anneal a heat treatment to which wafer is subjected in order to modify properties of materials/structures processed on its surface or in its bulk. *furnace, RTA*

anodic oxidation see *anodization.*

anodization anodic oxidation; process of oxidation in which growth of an oxide on the surface of a conductive solid immersed in the liquid electrolyte or in plasma containing oxygen is stimulated by the flow of electric current.

antimonides groups III-V semiconductor compounds of antimony (Sb), e.g. GaSb. InSb. *arsenides, nitrides, phosphides*

antimony, Sb group V element; *n*-type dopant (donor) in silicon; used mainly to dope epitaxial layers; diffusion coefficient in Si is comparable to arsenic and about order of magnitude lower than diffusion coefficient of phosphorous; forms antimonides with group III elements. *donor, antimonides*

anti-reflective coating, ARC thin layer of material deposited on the surface of the substrate to minimize reflection and to promote absorption of light in the substrate, e.g. in solar cells; also used to maximize absorption of the UV light in the layer of photoresist. *absorption*

antisite defect using GaAs as an example: Ga atom located on As site in GaAs crystal lattice, or As atom located on Ga site in GaAs crystal lattice. *defect*

APCVD see *Atmospheric Pressure CVD.*

APD see *Avalanche Photodiode.*

APM Ammonia hydroxide-hydrogen Peroxide-water Mixture; typically 0.25:1:5; same as *SC-1* and *RCA-1*; cleaning solution used primarily to remove particles from the surface; also capable of removing surface organics; strong solutions can etch/roughen silicon surface; forms chemical oxide (hydrophilic surface) on Si; applied at temperature between 40 °C and 70 °C; typically combined with megasonic agitation. *RCA clean, megasonic agitation, wet cleaning, hydrophilic surface, particles*

ARC see *anti-reflective coating.*

ArF excimer laser a chemical laser emitting 193 nm wavelength; used in photolithography; with various resolution enhancement techniques in place suitable for exposing geometries from 150 nm to 10 nm. *photolithography, enhancement techniques*

argon, Ar chemically inert gas; nontoxic; due to chemical inactivity and large mass of an ion used in semiconductor technology as an inert gas (more expensive than nitrogen) and in sputtering applications. *nitrogen, sputtering*

Arhenius plot allows determination of activation energy of the process from the slope of reaction rate constant vs. $1/T$ (T- temperature). *activation energy*

armchair GNR a type of graphene nanoribbon (GNR) which can display either semiconducting or metallic properties depending on the width of the nanoribbon; the term "armchair" refers to the specific configuration of carbon atoms at the edge of the nanoribbon. *graphene nanoribbon, zigzag GNR*

arsenic, As group V element; n-type dopant (donor) in silicon; features diffusion coefficient comparable to antimony and about order of magnitude lower than phosphorous; donor of choice in very small geometry devices; forms arsenides with group III elements. *donor, arsenides*

arsine, AsH₃ gaseous source of As for n-type doping of silicon by diffusion or ion implantation; highly toxic and flammable gas; must be handled with utmost care. *arsenic*

arsenides groups III-V semiconductor compounds of arsenic (As), e.g. GaAs, InAs. *antimonides, nitrides, phosphides*

ashing removal (by volatilization) of organic materials (e.g. photoresist) from the wafer surface using strongly oxidizing ambient; e.g. oxygen plasma.
barrel reactor, resist stripping

ASIC Application Specific Integrated Circuit; circuit designed and fabricated for a specific application; custom integrated circuit.

aspect ratio in semiconductor terminology taken as the ratio of the depth of the etched feature to its width on the surface normalized to 1; e.g. aspect ratio 10:1 defines the feature that is 10x deeper than wider.
trench

assembly process during which fully processed semiconductor device/ circuit in the form of a chip is mechanically and electrically connected to the package.
dicing, package, wire bonding

ATE Automated Test Equipment.

Atmospheric Pressure CVD, APCVD process of Chemical Vapor Deposition carried out at atmospheric pressure; typically results in the inferior film quality and conformality of coating as compared to Low Pressure CVD (LPCVD).
CVD, LPCVD, conformal coating

Atomic Force Microscopy, AFM a method used to visualize features of solid surfaces with near-atomic resolution; measurement of roughness of solid surfaces based on electrostatic interactions between surface and measuring tip; tip can be set above the surface, on the surface, or can tap the surface oscillating at high frequency (tapping mode).
surface roughness

Atomic Layer Deposition, ALD deposition method in which deposition of each atomic layer of material is controlled by a predeposited layer of precursor; precursors and various components of the film are introduced alternately; method features 100% step coverage and very good conformality; the method is commonly used in deposition of high-*k* dielectrics for MOS gates.
conformal coating, high-k dielectric, step coverage, CVD

9

Atomic Layer Epitaxy, ALE an atomic layer deposition (ALD) process which forms epitaxial layer.
Atomic Layer Deposition, epitaxy

Atomic Layer Etching, ALE subtractive process allowing selective, anisotropic removal of material (etching) with atomic layer precision; complements ALD processes in semiconductor nano-manufacturing.
Atomic Layer Deposition, selective etching, anisotropic etch

Auger electron an electron ejected from the solid as a result of two-stage ionization of atoms bombarded with high energy ions (Auger process); Auger electron carries energy specific to the atom from which it was ejected.
Auger Electron Spectroscopy

Auger Electron Spectroscopy, AES surface characterization and depth profiling method based on the determination of the energy of Auger electrons ejected from the surface bombarded with high-energy ions; only elements with atomic number above 2 can be detected by means of AES.
depth profiling

Auger recombination a band-to-band recombination process involving three particles interactions; energy released as a result of recombination is transferred to another electron or hole rather than released in the form of photon or phonon.
recombination, band-to-band recombination

autodoping dopant atoms evaporating from semiconductor surface region during high temperature treatments can be reintroduced into semiconductor causing undesired variations in dopant concentration at the surface; highly undesired effect; of particular concern in high-temperature epitaxial deposition processes.
epitaxy, outdiffusion

avalanche breakdown breakdown caused by avalanche multiplication of charge carriers in the space charge region of the *p-n* junction at the very high reverse-bias voltage; results in rapid increase of reverse current across the junction; the most common breakdown mechanism in *p-n* junction (other breakdown mechanism is based on the Zener effect).
breakdown Zener effect, avalanche diode

avalanche diode a *p-n* junction diode designed to allow an avalanche breakdown without sustaining permanent damage.
reversible breakdown, soft breakdown

avalanche ionization see *avalanche multiplication.*

avalanche multiplication generation of electron-hole pairs due to impact ionization in the depleted region of semiconductor (e.g. in reverse-biased *p-n* junction) which causes continuation of the impact ionization events, and thus, further increases the number of carriers.
depleted region, p-n junction, multiplication factor

avalanche photodetector see *avalanche photodiode.*

avalanche photodiode, APD a photodiode in which response to the illumination is enhanced by the avalanche multiplication process.
avalanche multiplication, photodiode, photoelectric effect

B

back contact the electrical contact to the back surface of the wafer.
ohmic contact

Back-End-of-Line processes, BEOL operations performed on the semiconductor wafer in course of the chip manufacturing after the first metallization process.
Front-End-of-Line processes

backscattering a redirection of light (electromagnetic wave) or electrons penetrating semiconductor back to the direction from which they came; occurs due to the scattering as opposed to reflection.
scattering

Ball Grid Array, BGA, packaging IC chip assembly and packaging technology; chip is flipped over and connected to the package via an array of solder balls positioned to match metal pads on the surface of the chip; compatible with surface mount technology.
assembly, package, flip chip technology, surface mount technology

ballistic transport   a motion of electrons in highly confined semi-conductor structures in the presence of very high electric field with velocity much higher than their thermal equilibrium velocity; ballistic electrons behave as waves and are not subjected to scattering, and hence, can move with ultra-high velocity; allows ultra-fast devices.
scattering, thermal equilibrium, particle-wave duality

ballroom cleanroom   a cleanroom layout featuring one common space shared by the technical personnel, tools and some service functions; no designated and isolated areas; relatively straightforward and less expensive per sq. ft. than other advanced cleanroom designs, but not effective in assuring cleanroom class lower than 1,000.
cleanroom, cleanroom class

band alignment   see *band-edge lineup*.

band-edge lineup   a lineup of the edges of conduction and valance bands of two different semiconductors forming an abrupt heterojunction; three types of lineup include straddling, staggered, and broken gap.
abrupt heterojunction, broken gap, staggered, straddling

band offset   a measure of "misalignment" between energy levels at the interface between two solids; e.g. Si and gate dielectric in MOS structures.
energy levels

band-to-band generation   the electron-hole pair generation as a result of an electron gaining energy (through increased temperature or illumination) and transitioning from the valance band to the conduction band of semiconductor.
generation, electron-hole pair, thermal generation

band-to-band recombination   the electron-hole pair recombination as a result of electron loosing (releasing) energy and transitioning from the conduction band to the valance band of semiconductor.
recombination, radiative recombination

band-to-band tunneling   see *Zener effect*.

bandgap   see *energy gap*.

bandgap engineering   processes by which bandgap of semiconductor is altered in the desired way by changing its chemical composition; e.g. by adding Al the bandgap of GaAs can be altered from 1.424 eV to 2.026 eV; used in superlattice fabrication for laser diodes and transistors.
energy gap, superlattice

BARC   see *bottom anti-reflective coating.*

BARITT   Barrier Injection Transit-Time diode; microwave diode.

barium strontium titanate, (BaSr)TiO$_3$, BST   dielectric featuring very high k (in the range 160-600) at thicknesses exceeding about 40 nm; displays ferroelectric properties; used in storage capacitors, can be deposited by the variety of methods including MOCVD, sputtering, and misted deposition (LSMCD).
MOCVD, sputtering, mist deposition

barrel reactor   the "barrel" shaped reactor most commonly used for batch plasma resist stripping.
ashing, batch process, resist stripping

barrier height [eV]   the height of the potential barrier at the junction/ interface (e.g. *p-n* junction or metal-semiconductor contact) reflecting the difference between potential at the surface and potential in the bulk of semiconductor; affected by the type of material/ambient with which semiconductor is in contact.
potential barrier

barrier metal   thin layer of metal, e.g. TiN, formed between other metal and semiconductor (or insulator) to prevent potentially harmful inter-actions between these two materials, e.g. Al spiking of silicon.
Al contact, spiking, titanium nitride

base   a region in bipolar transistor sandwiched between emitter and collector; typically *p*-type so that highly mobile electrons act as minority carriers in the base; should be very thin to allow rapid transfer of minority carriers from emitter to collector; electric field is created in the base by non-uniform doping to accelerate minority carriers moving from emitter to collector.
base width, bipolar transistor, collector, emitter

base punchthrough occurs in bipolar transistors when at the high collector-base voltage the collector space charge region expands over the entire base width and comes in contact with the emitter space charge region; in such situation transistor action is no longer possible.
base, base width, space charge region

base pushout the Kirk effect; apparent base-width increase in a bipolar transistor caused by the high concentration of carriers being swept from the base to the collector.
base width

base transport factor defines efficiency of the minority carrier transport through the base of a bipolar transistor, i.e. from the emitter junction to the collector junction; should be as close to 1 as possible; to a significant degree determined by the base width.
base width, bipolar transistor

base width distance between metallurgical emitter and collector junctions in bipolar transistor; key geometrical characteristic of the bipolar transistor defining its performance; narrower the base, more efficient transfer of minority carriers from the emitter to collector across the base, and hence, better performing transistor.
base, metallurgical junction

batch process the process in which several wafers are processed at the same time; opposite to a single wafer process; e.g. thermal oxidation of silicon wafers in the furnace is an example of the batch process.
furnace, single wafer process

batch reactor the reactor in which several wafers can be processed at the same time; e.g. oxidation furnace or wet bench.
batch process

BBUL packaging Bumpless Build-Up Layer technology, BBUL "grows" the package around the silicon die resulting in the thinner, higher-performance, consuming less power processors; this is in contrast to the practice of manufacturing the processor die separately and later bonding it to the package.
package

14

BCCD Buried Channel Charge-Coupled Device.
Charge Coupled Devices

beam geometrically confined, directional stream of electromagnetic radiation or electrically charged particles such as electrons or ions.
electron beam, ion beam

BEOL see *Back-End-of-Line.*

BGA see *Ball Grid Array.*

BHF buffered hydrofluoric acid; see *buffered oxide etch.*

bias-temperature instability an adverse effect in MOS devices unraveled by means of the *bias-temperature stress* test.

bias-temperature stress, BTS testing method in the MOS device characterization in which MOS capacitor biased in accumulation is maintained over the period of time at the increased wafer temperature (typically not exceeding 100 °C); a test devised to unravel the time- and bias dependent variations in the characteristics of the MOS gate; e.g. related to interface traps and mobile ions.
accumulation, NBTI, interface trap, mobile ion

biaxial strain occurs when the crystal has stress acting in its lattice in two directions (*x-y*) along its surface; also referred to as a global strain; just like in the case of an uniaxial strain, when created in the MOSFET channel its effect is different for NMOSFET and PMOSFET and depends on whether stress is tensile or compressive.
strain, uniaxial strain, tensile strain, compressive strain

BiCMOS Bipolar CMOS; technology which combines bipolar and CMOS devices on the same chip; significant gains in circuit functionality outweigh increased process complexity.
CMOS, bipolar transistor

binary semiconductor semiconductor compound consisting of two elements; e.g. SiC, GaAs, CdS.
elemental, ternary, quaternary semiconductor

bipolar device a semiconductor device operation of which is based on the use of both majority and minority charge carriers; all *p-n* junction based devices fall into this category; opposite to an unipolar device.
unipolar device

bipolar junction transistor, BJT a *p-n* based junction transistor consisting of three semiconductor regions (emitter, base, and collector) with alternating conductivity type (i.e. *n-p-n* or *p-n-p*); current flow comprises both majority and minority carriers, hence, "bipolar"; BJT's performance and characteristics are to a significant degree dependent on its layout (vertical, lateral) and geometry (base width); BJT is a current controlled transistor; implemented using silicon homojunctions or multi-material heterojunctions.
transistor, p-n junction, lateral transistor, unipolar transistor, HBT

"bird beak" sharp, bird beak-like shaped geometrical feature at the Si surface at the edge of LOCOS oxides; undesired as it may cause reliability problems in CMOS devices.
LOCOS

bit the smallest unit of information in computer language (the smallest amount of digital information); represented by "*1*" and "*0*" which in various combinations represent characters and numbers; digits "*1*" and "*0*" represent two states in binary notation.

BITS Burn-In Test Socket.

BJT see *bipolar junction transistor.*

black phosphorus an allotrope of phosphorus which in 2D configuration displays excellent semiconductor properties; see *phosphorene.*

blank a part of the mask used in photolithography which is transparent to UV light; quartz is commonly used as a blank in photomasks.
opaque material, photomask, quartz

BNDC Bulk Negative Differential Conductivity.

boat *(i)* a device designed to hold semiconductor wafer during thermal processes; made of high purity temperature resistant materials such as fused silica, quartz, polycrystalline Si, or SiC, *(ii)* a device designed to

contain source material during thermal evaporation acting at the same time as a heater; made of temperature resistant highly conductive material (e.g. tungsten) through which current is passed.
evaporation, thermal oxidation

Body-Centered Cubic, BCC, cell   a sub-lattice in cubic system (along with FCC and simple cubic lattice); formed by placing a lattice point at the center of the simple cube.
crystal lattice, cubic system, Face-Centered Cubic

BOE   see *Buffered Oxide Etch*.

Bohr exciton radius   a distance between the electron and the hole forming an exciton.
exciton

Boltzmann constant $k = 1.38 \times 10^{-23}$ J/K

Boltzmann transport equation   a fundamental relationship describing transport of free carriers in semiconductors.

bonded SOI   the SOI substrates formed by bonding two silicon wafers with oxidized surfaces; following bonding one wafer is polished down to the desired thickness of active layer with interface oxide becoming a buried oxide, alternative approach in bonded SOI technology: *smart cut®*
wafer bonding, cleaved SOI, SIMOX, smart cut®

bonded wafer   see *wafer bonding*.

Boro-Phospho-Silicate Glass, BPSG   silicon dioxide (silica) with boron and phosphorus added to lower temperature at which glass (oxide) starts to flow from about 950 °C for pure SiO_2 to about 500 °C for BPSG; used to planarize the surface; typically deposited by CVD.
planarization, PSG

boron   an element from the group III of the periodic table acting as an acceptor (*p*-type dopant) in silicon; the only *p*-type dopant broadly used in silicon device manufacturing.
acceptor, p-type dopant

boron nitride, BN the widest bandgap material displaying semiconductor properties (6.4 eV in the case of cubic c-BN); intrinsic bandgap nature (direct or indirect) is not firmly established because of the insufficient quality of BN crystals available; features very high thermal and chemical stability.
aluminum nitride

boron penetration term usually refers to penetration of the gate oxide by boron from heavily *p*-doped poly-Si gate contact in PMOS part of the CMOS cell; at elevated temperature boron from the gate contact readily segregates into the adjacent oxide causing reliability problems.
ONO, segregation coefficient

Bosch process variation of the DRIE process assuring anisotropy of the very deep etches involved in MEMS and TSV processing.
Deep Reactive Ion Etching, anisotropic etch, MEMS, TSV

Bose-Einstein distribution function (d.f) a probability function which describes distribution of electron states in solids.
Fermi-Dirac d.f, Maxwell-Boltzman d.f

bottom anti-reflective coating, BARC a layer of material deposited underneath photoresist to prevent reflection from the wafer surface of the light passing through the layer of photoresist; in photolithography used to enhance control of critical dimensions (CD) by suppressing standing waves effects and reflective notching in photoresist.
anti-reflective coating, standing waves effect, reflective notching

bottom-gate TFT a Thin-Film Transistor with gate contact formed on the insulating substrate first and then covered with gate dielectric and a layer of thin-film semiconductor.
Thin-Film Transistor, top-gate TFT

bottom-up process a way of building functional structures on semiconductor substrates which relies on the self-assembly of molecules in the previously patterned areas.
top-down process, surface functionalization

boundary layer a layer in contact with the solid surfaces immersed in gaseous or liquid medium; within the boundary layer characteristics of the medium are different than in its bulk.

BOX see *buried oxide*.

BPSG see *Boro-Phospho-Silicate Glass*.

breakdown highly undesired, damaging effect occurring in the presence of high electric field and causing originally high-resistance element (e.g. MOS capacitor or reverse biased *p-n* junction) to allow uncontrolled flow of current; typically, an irreversible effect permanently damaging semiconductor element; also occurs in homogenous materials in the presence of the very high electric field.
oxide breakdown

breakdown field electric field at which breakdown in the material/ device occurs and material/device looses its resistance against current flow; breakdown field in Si is 2.5×10^5 V/cm, GaAs 3×10^5 V/cm, 6H-SiC 4×10^6 V/cm, thermal SiO_2 10^7 V/cm.

breakdown, hard see *hard breakdown*.

breakdown, soft see *soft breakdown*.

breakdown voltage the voltage at which breakdown in any given semiconductor device occurs and current increases uncontrollably (unless limited by the external circuit).
ramp voltage oxide breakdown

Bridgman growth (method) a method used to grow single-crystal compound semiconductors, typically III-V, e.g. GaAs; uses multi-zone furnace in which Ga and As are contained in the ampoule and in contact with GaAs single-crystal seed; the GaAs melt is passed from higher to lower temperature zone; conceptually similar to float-zone (FZ) crystal growth method.
Float-Zone crystal growth, Czochralski crystal growth

Brillouin zone the space which is limiting wave vector *k* for electrons in the crystal.

broken gap a lineup of the edges of conduction and valance bands in two different semiconductors forming an abrupt heterojunction such that the edge of the valence band in one semiconductor forming

heterojunction is below the edge of the conduction band in the other semiconductor.
band-edge lineup, abrupt heterojunction, staggered gap, straddling gap

BRT Base Resistance controlled Thyristor.
thyristor

brush scrubbing cleaning of semiconductor surfaces using rotating brushes; used to remove heavy residues which cannot be efficiently removed without mechanical interactions; used in post-CMP wafer cleaning.
post-CMP cleaning, scrubbing

BST see *barium strontium titanate, (BaSr)TiO₃.*

BTS see *bias-temperature stress.*

buffer layer term typically refers to a layer sandwiched in between two single-crystal materials to accommodate difference in their crystallographic structures (lattice constants).
graded junction, superlattice, heteroepitaxy

buffered oxide etch BOE the HF-water solution with ammonium fluoride, NH_4F, added to prevent depletion of fluorine during oxide etching; at 34% NH_4F : 6.8% HF : 58.6% H_2O BEO etches thermal SiO_2 at the rate of 100 nm/min.
hydrofluoric acid, etching

bulk part of the semiconductor material away from the surface where its physical properties are the same in all three directions; at the absence of electric field remains electrically neutral (as opposed to the surface of the same material).
surface

bulk CMOS the CMOS circuitry (cell) implemented in a standard bulk Si wafer rather than in a thin layer of Si on insulator (SOI substrate).
CMOS, SOI wafer

bulk properties physical properties of semiconductor material away from the surface; differ significantly from the same properties at the

surface, in the near-surface region, or in the same material in the form of thin-film.
bulk, surface properties, thin-film

bulk recombination recombination of the free charge carriers in the bulk of semiconductor; occurs primarily by band-to-band recombination.
recombination, band-to-band recombination, surface recombination

bulk wafer the homogenous semiconductor wafer with no additional distinct layer formed on its surface.
wafer, SOI wafer, epitaxial extension, bulk properties

buried oxide, BOX the oxide layer in SOI substrates (wafers); oxide (SiO_2) is buried in silicon wafer at the depth ranging from less than 100 nm to several micrometers from the wafer surface depending on application; formed by oxidation or oxygen implantation depending on the type of SOI substrate; thickness of BOX is typically in the range from about 40 nm to about 100 nm.
bonded SOI, SIMOX

burn-in reliability testing procedure used to force defective devices to fail; process designed to detect early failures of semiconductor devices; device is subjected to electric stress at elevated temperature for a lengthy period of time.

burst noise see *Random Telegraph Noise*.

C

cadmium mercury telluride, CdHgTe ternary narrow-bandgap semiconductor material broadly used in the manufacture of infrared photodetectors.
ternary semiconductor, photodetector

cadmium selenide, CdSe II-VI group semiconductor; bandgap $E_g = 1.74$ eV, direct; wurzite crystal structure; useful in various photonic device and solar cell applications; also explored in the form of nanocrystalline quantum dots for LED applications.
nanocrystalline quantum dot, solar cell

cadmium sulfide, CdS II-VI group semiconductor; bandgap E_g = 2.42 eV, direct; wurzite crystal structure; used primarily in light detection applications typically in the form of a photoresistor; also used in optics for waveguides and beam splitters.
photoresistor

cadmium telluride, CdTe II-VI group semiconductor; bandgap E_g = 1.56 eV, direct; zinc blend crystal structure; with energy gap matching solar spectrum very useful in solar cell applications; alloyed with mercury makes an infrared detector material CdHgTe.
thin-film solar cell

CAIBE Chemically Assisted Ion Beam Etching; combines features of chemical and physical etching.
chemical etching, physical etching

calcium, Ca common contaminant in semiconductor device manufacturing; originates primarily from chemicals and water; roughens Si surface during thermal oxidation; has an adverse effect on the gate oxide reliability.
surface roughness, contaminant

cantilever loading loading of the wafers into a horizontal furnace in such way that the boat supporting wafers is not coming in contact with inside walls of the process tube, thus, avoiding generation of particles.
horizontal furnace

capacitance-voltage, C-V, measurements several parameters of semiconductor materials and material systems can be determined by measuring C-V characteristics of designated test devices; routinely used for process diagnostics and monitoring in MOS technology; allows determination of the interface trap density, oxide fixed charge, oxide total charge, as well as flat-band voltage.
flat-band voltage, oxide fixed charge, interface trap

capping layer protective layer formed to prevent changes in the material properties during processing; e.g. a layer on the top surface of the porous low-k dielectric that is sealing off pores in its structure.
porous dielectric, low-k dielectric

capture cross-section effective area of a trap; defines capability of the trap to capture free charge carrier.
trap

CAR see *Chemically Amplified Resist.*

carbon, C an element in the same group IV of the periodic table as silicon and germanium; among its allotropes diamond (a single-crystal carbon) is an excellent wide-bandgap ($E_g = 5.45$ eV) semiconductor while graphite features very high electrical conductivity, but weak semiconductor properties; one-atom thick sheets of carbon either planar (graphene) or rolled to form a tube (carbon nanotube) display a range of excellent physical properties; as a part of the compound with silicon carbon forms silicon carbide, SiC, which is a key wide-bandgap semiconductor.
diamond, graphene, carbon nanotube, silicon carbide

carbon contamination common dry etching chemistries of semiconductors contain carbon which ends up being implanted during Reactive Ion Etching processes in the near-surface (~ 3 nm) region of etched material.
Reactive Ion Etching, RIE damage

carbon doped oxide, CDO an oxide to which carbon is added to reduce its dielectric constant k; used as a low-k interlayer dielectric in advanced interconnect schemes.
low-k dielectric, inter-layer dielectric

carbon electronics carbon appears in nature in multiple forms each displaying distinct electronic properties; with time the role of carbon in the advancement of electronics will only grow.
carbon, diamond, graphene, carbon nanotube, fullerens, silicon carbide

carbon nanotube, CNT a rolled sheet of hexagonal carbon structured in the "chicken wire"-like form; essentially a seamless tube of a single layer of carbon (graphene) in the single-walled (SWCNT) or double-walled (DWCNT) configuration; part of the fullerene structural family; depending on the structure displays either semiconductor or metallic properties, hence, can be used in the range of applications.
fullerenes, single-walled nanotube, multi-walled nanotube

"Caro" clean   same as *SPM clean.*

carrier   in semiconductor terminology term "carrier" is synonymous with the term "charge carrier".
charge carrier, electron, hole

carrier crowding   see *current crowding.*

carrier extraction   process of removing minority carriers from the semiconductor region that remains under the influence of an electric field; e.g. in a reverse-biased *p-n* junction electrons are extracted from the *p*-type region and holes are extracted from the *n*-type region.
minority carrier

carrier generation   see *generation.*

carrier injection   the process of introducing charge carriers from one region within semiconductor device to another, e.g. in a forward-biased *p-n* junction electrons are injected into the *p*-type region and holes are injected into the *n*-type region.
emitter

carrier recombination   see *recombination.*

catastrophic breakdown   an irreversible breakdown resulting in permanent inability of the device or material to control current flow.
hard breakdown, breakdown

cavity   see *optical cavity.*

CBE   Chemical Beam Epitaxy.
epitaxy

CCD   see *charge-coupled devices.*

CCS   see *constant-current stress.*

CD   see *critical dimension.*

CDO   see *carbon doped oxide.*

CEL see *Contrast Enhancement Layer*.

ceramic package an electronic/photonic/MEMS device housing package made out of ceramic material; due to the high modulus of elasticity and low coefficient of thermal expansion ceramic packages find a broad range of applications.
package

CERDIP Ceramic Dual-In-line Package.
DIP, package

CFM Contamination Free Manufacturing.

cfm cubic feet per minute, see *pumping speed*.

chain scission a process of bonds breaking in polymeric chains as a result of illumination; foundation of a positive resist technology in photolithography; bond breaking (chain scission) makes originally insoluble positive resist soluble in in the developer.
crosslinking, positive resist

chalcogenides chemical compounds involving chalcogens; display physical properties of interest in semiconductor device engineering including solar cells; e.g. binary GeS_2, ternary $Ge_2Sb_2Te_5$, or quaternary $CuIn_xGa_{(1-x)}Se_2$.
chalcogens

chalcogenides CVD a Chemical Vapor Deposition of chalcogenides thin films.
CVD

chalcogens elements in group 16 of the periodic table among which sulfur, selenium and tellurium form compounds used in thin-film deposition processes.
CVD

channel high conductivity region connecting source and drain in Field Effect Transistors; electrical conductivity of the channel can be changed by changing gate potential; by turning channel "on" and "off" the switching action in FETs is accomplished.
Field Effect Transistor, drain, source

25

channel length distance between source and drain in the Field Effect Transistors (FET); shorter the channel faster switching by the FET can be achieved; reduction of the channel length in MOSFETs is a driving force behind the progress in semiconductors electronics; strictly speaking not the same, but in common usage synonymous with the term "gate length".
channel, gate length, short-channel effects, gate scaling

channel stop p^+ implanted layer underneath oxide pad in LOCOS isolation scheme in CMOS devices; put in place to prevent formation of an inversion layer which would create a conductive channel between PMOS and NMOS parts of the CMOS cell.
latch-up, LOCO, STI

channeling the effect occurring during ion implantation into crystalline solids; implanted specie may enter an open "channel" in the crystal lattice as a result of which it may penetrate the solid deeper than other implanted species subjected to collisions with atoms in the lattice; probability of channeling increases with implantation energy; its impact is reduced by adjusting implantation angle.
ion implantation, implantation energy, implantation angle

charge carrier a particle carrying an electric charge which is free to move in the conductive solid; electrons in conductors, electrons and hole in semiconductors.
electron, hole, ion

Charge Coupled Device, CCD an imaging device based on a silicon MOS capacitor structure; exploits sensitivity of charge distribution in the space-charge region of MOS capacitor to illumination and converts optical signal into electrical signal; commonly used as an imaging device in video equipment; manufactured in very densely packed arrays in numbers per chip defining number of "pixels".
space-charge region, CMOS image sensor

charge screening an effect which reduces strength of interactions between electric charges by reducing electrostatic field.

charge, static see *static charge.*

charge-to-breakdown, Q_{bd} a measure of the reliability of thin gate oxides in MOS devices; Time Dependent Dielectric Breakdown (TDDB) technique; current of known density is forced through the oxide at the constant gate voltage; point in time at which voltage drops (indicating oxide breakdown) is determined; knowing current density and time to breakdown, total charge needed to break the oxide (Q_{bd}) is determined.
GOI, oxide breakdown, ramp voltage oxide breakdown, TDDB

chemical etching process of either wet (liquid-phase) or dry (gas-phase) etching through the chemical reaction between chemically reactive etching species and etched material; chemical etching is isotropic and selective.
etch, isotropic etch, selective etch, physical etching, wet etching, dry etching

Chemical-Mechanical Planarization, CMP increased force mechanical interactions between polishing pads and the patterned wafer surface enhanced by the chemically reactive slurry; a combination of chemical etching and abrasive polishing carried out for the purpose of surface planarization through the removal of excess material; commonly used in the definition of metal interconnect pattern; key process in back-end of line IC manufacturing.
surface planarization, BEOL, damascene

Chemical Mechanical Polishing, CMP a reduced-force mechanical interactions between polishing pads and the wafer surface enhanced by the chemically reactive slurry; a combination of chemical etching and abrasive polishing carried out for the purpose of reducing surface roughness; can be performed on the front surface of the wafer or front and back surfaces simultaneously.
wafer fabrication, wafer thinning, slurry, double-sided polishing, surface planarization

chemical oxide an oxide growing on the surface of semiconductors during wet cleaning and rinsing operations; highly hydrated with composition departing from stoichiometric native oxide; in the case of Si chemical oxide can be defined as SiO_x with $O < x < 2$.
APM, native oxide

Chemical Vapor Deposition, CVD the most common thin-film deposition method in advanced semiconductor manufacturing; deposited species are formed as a result of the chemical reaction between gaseous reactants at elevated temperature in the vicinity of the substrate or on its surface; solid product of the reaction is deposited on the surface of the substrate; used to deposit films of semiconductors (crystalline and non-crystalline), insulators as well as metals; key variations of CVD processes include Atmospheric Pressure CVD (APCVD), Low Pressure CVD (LPCVD), Plasma Enhanced CVD (EPCVD), Metal-Organic CVD (MOCVD), Atomic Layer CVD (ALCVD) *(definition of each is included in this glossary)*.
Physical Vapor Deposition, Physical Liquid Deposition

Chemically Amplified Resist, CAR a photoresist with photoactive component added to stimulate photochemical process during resist exposure; an advanced photoresist needed in deep UV photolithography.
enhancement techniques, deep UV

chemisorption adsorption of species (adsorbates) at the solid surface by formation of chemical bond between the adsorbate and the surface.
physisorption

chip a part of semiconductor wafer containing the entire circuit, also referred to as a "die".
die, integrated circuit

chrome mask a photolithographic mask (photomask) which uses thin film of chrome as an opaque material; due to sharper edges, mechanical durability and high optical density of Cr the chrome masks offer superior performance as compared to photomasks using emulsion as opaque material.
opaque material, photoemulsion mask, photolithography, photolithographic mask

CIGS copper indium gallium (di)selenide; chemical formula: $CuIn_xGa_{(1-x)}Se_2$; used as a light-absorbing polycrystalline thin-film material system in efficient (above 19%) heterojunction solar cells.
solar cell, chalcogenides

CIM Computer Integrated Manufacturing.

CIS copper indium diselenide ($CuInSe_2$); in the form of thin-film of interest in solar cell engineering.
solar cell, chalcogenides

cleaning the process of removing contaminants (particles as well as metallic and organic contaminants) from the surface of the wafer; can be implemented using liquid chemicals (wet cleaning) of gases (dry cleaning); the most frequently applied processing step in traditional semiconductor device manufacturing sequence; an element of the broadly understood surface engineering; implemented differently in bulk wafer and thin-film based device technologies; standard cleaning techniques are not compatible with processes involved in thin-film device processing, organic semiconductors, and nanomaterial device systems.
contaminant, dry cleaning, wet cleaning, surface engineering

cleanroom an enclosed ultra-clean space in which semiconductor manufacturing takes place; features counts of airborne particles reduced to technically feasible minimum and strictly controlled temperature and humidity of ambient air.
particles, ballroom cleanroom, HEPA filters, ULPA filters

cleanroom class a number indicating class (quality) of a clean room; defined in terms of the number of particles of given size per cubic foot or cubic meter of air; e.g. (for the sake of concept illustration only) Class 10: maximum 10 particles 0.5 μm per ft^3; Class 1: maximum 1 particle 0.5 μm per ft^3.
cleanroom

cleaved SOI process used to fabricate bonded SOI substrates; before bonding one wafer is implanted with hydrogen to the depth which will determine thickness of an active layer in future SOI wafer; following bonding wafer is subjected to an anneal during which implanted wafer splits along the plane stressed with implanted hydrogen - one section, very thin, remains bonded to the other wafer forming SOI substrate while the other can be re-used in the fabrication of additional SOI substrates; process is also know as a Smart Cut® (registered name SOITEC S.A.)
bonded SOI, active layer, cleaving

cleaving splitting of semiconductor crystal along crystallographic planes or induced in the crystal lattice strained regions; typically initiated by rapid thermal treatment of the crystal.
crystal plane, strain

cluster, linear the cluster tool in which process modules are installed "in-line", i.e. wafers can only be processed in sequence in which process modules are integrated.
process integration, cluster tool

cluster, radial the cluster tool in which process modules are installed on the sides of the centrally located wafer handler platform; single robotic arm installed in the center transfers wafers in between the modules in the flexible sequence.
process integration, cluster tool

cluster tool a tool comprising two or more mechanically and software integrated process modules installed on the central wafer handler platform; allows several operations to be performed on the wafer without exposing it to ambient air.
integrated processing, handler platform

CMOS Complementary Metal Oxide Semiconductor structure; consists of N-channel and P-channel MOS transistors (MOSFETs); due to the very low power consumption and dissipation, very low energy needed to perform switching as well very low standby current the CMOS is a very effective cell for the implementation of digital functions; CMOS is by far the most important device configuration in state-of-the-art silicon digital integrated circuits.
digital integrated circuit, MOSFET, NMOSFET, PMOSFET

CMOS image sensor an alternative to CCD design of image sensors converting optical image into electric signal; manufactured using CMOS process; choice between CCD and CMOS image sensors depends on specific application.
charge-coupled device, CMOS

CMOS inverter a pair of two complementary MOS transistors (P-channel and N-channel) with the source of the N-channel transistor connected to the drain of the P-channel transistor and the gates

connected to each other; the output (drain of the *P*-channel transistor) level is high whenever the input (gate) level is low, and the other way round; CMOS inverter is a basic building block of CMOS digital circuits.
CMOS, logic gate, NMOSFET, PMOSFET

CMP   see *Chemical Mechanical Polishing.*

CMP   see *Chemical Mechanical Planarization.*

CNFET   Carbon Nanotube Field Effect Transistor; a MOSFET using carbon nanotube as a channel.
carbon nanotube, MOSFET

CNT   see *carbon nanotube.*

cobalt silicide, CoSi$_2$ contact (ohmic) material in Si technology; resistivity 16-20 $\mu\Omega$-cm; formed at the sintering temperature of 900 °C.
ohmic contact, silicide, sintering temperature TiSi$_2$, WSi$_2$

cold cathode   see *field emission.*

coldwall reactor   thermal reactor using inductive heating; energy is coupled primarily to the wafer holder while walls of the process chamber remain at the low temperature.
hotwall reactor, radiant heating

collector   a region in bipolar transistor collecting carriers from the base; should be lightly doped to assure proper transistor operation; in an integrated version very highly doped sub-collector region is formed to assure ohmic characteristics of the collector contact.
bipolar transistor, sub-collector contact, ohmic contact, common collector configuration

colloidal semiconductor   a colloidal solution in which dispersed phase consists of nanoparticles of semiconductor material; using spin-on deposition, mist deposition or other Physical Liquid Deposition (PLD) methods colloidal semiconductor can be converted into a solid thin-film.
colloidal solution, Physical Liquid Deposition

colloidal solution   a solution in which *nm*-sized insoluble particles are dispersed in the liquid medium; by definition must involve two distinct

phases: a dispersed phase (the suspended particles) and a continuous phase (the medium of suspension); in semiconductor engineering used as precursors in nanocrystalline quantum dots films formation.
colloidal semiconductor, nanocrystalline quantum dot

common-base configuration a circuit (amplifier) configuration of a bipolar transistor where emitter is the input electrode, collector is the output electrode and base is the reference electrode for both input and output.
common-emitter configuration, bipolar transistor

common-base current gain the ratio of the collector current to the emitter current in bipolar transistors in common-base configuration.
bipolar transistor

common-base cut-off frequency frequency at which current gain in the common-base configuration of the bipolar transistor drops to 1.

common-collector configuration circuit (amplifier) configuration of a bipolar transistor where base is the input electrode, emitter is the output electrode and collector is the reference electrode for both input and output.
common-emitter configuration

common-collector current gain the ratio of the emitter current to the base current in bipolar transistors in common-collector configuration.
bipolar transistor

common-emitter configuration circuit (amplifier) configuration of a bipolar transistor where base is the input electrode, collector is the output electrode and emitter is the reference electrode for both input and output.
common-base configuration

common-emitter current gain the ratio of the collector current to the base current in bipolar transistor in common emitter configuration.
transistor gain

common-emitter cut-off frequency frequency at which current gain in the common-emitter configuration in the bipolar transistor drops to 1.

compensated semiconductor   a semiconductor   which contains donor (concentration N_D) and acceptor (concentration N_A) dopant atoms in the same region; in n-type compensated semiconductor $N_D > N_A$; in p-type compensated semiconductor $N_D < N_A$.
dopant

complementary MOS   see *CMOS.*

complex oxides   a family of ABO_x oxides (e.g. $SrTiO_3$) displaying unusual, potentially very useful physical properties due to which they can be engineered to behave as dielectric, magnetic or superconducting materials.
functional oxides, strontium titanate

compound semiconductors   semiconductors formed using two or more elements; compound semiconductors do not appear in nature; they are synthesized using primarily elements from groups II through VI of the periodic table; distinct classes of compound semiconductors include: AII-BVI (e.g. CdSe), AIII-BV (e.g. GaAs) and AIV-BIV (e.g. SiC).
semiconductors, AII-BVI, AIII-BV, AIV-BIV, elemental semiconductors, inorganic semiconductors

compressive stress   stress resulting in compression of the material; tensile stress acts in the opposite direction; stress is a common occurrence in semiconductor structures at the interface between crystals featuring different lattice constants.
stress, tensile stress, lattice mismatch

computational photolithography   an approach to the improvement of the resolution of the photolithographic pattern transfer process based on computer models predicting impact of diffraction, interference and other resolution-limiting effects and using these models to optimize masks design.
enhancement techniques, mask, OPC

concentrated photovoltaics   lenses or other optical means are used to concentrate a large amount of solar energy (sunlight) onto small area photovoltaic cells, thereby, allowing generation of the same amount of electricity using much smaller (reduced area) solar panels.
solar cell, photovoltaic effect

conduction band   the upper energy band in semiconductor separated by the energy gap (bandgap) from the valance band; conduction band is not completely filled with electrons, hence, electrons can conduct in the conduction band; in metals valence band and conduction band overlap.
energy gap, valence band

conductivity, electrical   reciprocal of resistivity; unit S (Siemens) = $1/\Omega$; measure of semiconductor ability to conduct current; depends on dopant concentration and mobility of charge carriers.
resistivity, mobility

conformal coating   a coating (deposited film) which thickness remains the same regardless of the underlying geometrical features of the substrate surface.
shadowing effect

constant-current stress, CCS   technique used to study time-dependent breakdown of the gate oxide; constant current featuring predetermined density is injected into the oxide until it breaks downs (gate voltage drops to zero).
gate oxide, GOI, TDDB

constant-source diffusion   also known as unlimited-source diffusion or pre-deposition; concentration of diffusant (dopant) on the surface of the wafer remains constant during the diffusion process, i.e. while some dopant atoms diffuse into the substrate additional dopant atoms are continuously supplied to the surface of the wafer.
limited-source diffusion, pre-deposition, redistribution

constant-voltage stress, CVS   technique used to study time-dependent breakdown of the gate oxide; constant voltage is applied to the gate until oxide breaks down (rapid, irreversible increase of the gate current).
gate oxide, GOI, TDDB

contact angle   see *wetting angle*.

contact printing   printing mode used in photolithography in which during exposure photomask and the wafer are in physical contact; allows resolution below 1 μm depending on the wavelength of exposing light,

but may cause mask damage; in advanced nanoelectronic manufacturing replaced with projection printing (steppers).
projection printing, proximity printing

contact resistance resistance of the metal-semiconductor contact commonly resulting from the presence of an ultra-thin oxide and/or organic contaminants at the interface
metal-semiconductor contact, chemical oxide, organic contaminant

contaminant an alien element or specie from the medium (e.g. from process gases, liquids, clean room air, etc.) in contact with semi-conductor; contaminant may adhere to its surface and adversely affect the process and eventually performance of the processed device; the most common types of contaminants in semiconductor processing: particles, organic contaminants and metallic contaminants each having a distinct adverse effect on the process and/or device performance.
particle, metallic contaminant, organic contaminant

contamination presence of contaminants in the process ambient and/or on the wafer surface.
contaminant

contamination control prevention of penetration of process ambient by contaminants.
cleanroom

continuity equation an equation that states that the current density in semiconductor may vary with time only due to generation-recombination processes.
generation, recombination

Contrast Enhancement Layer, CEL a layer in the photoresist stack added to increases resolution of the pattern transfer process.
enhancement techniques

conversion efficiency in solar cell terminology the ratio of the output electrical power to the power of incident solar radiation; expressed in %; varies from few % to over 40% depending on the type of solar cell.
solar cell

COO   see *Cost of Ownership*.

COP   see *Crystal Originated Pits* and *Crystal Originated Particle*.

copper, Cu *(i)* metal of choice for interconnects in advanced ICs; features second lowest resistivity among metals (1.7 $\mu\Omega$-cm); advantages over aluminum: no electromigration and lower resistivity; *(ii)* defect causing contaminant if allowed to penetrate silicon; very fast diffusant in silicon; results in the reduced lifetime of minority carriers by defect decoration.
electromigration, metallic contaminants, defect decoration

copper interconnect an interconnect scheme in advanced integrated circuits using copper as a metal for interconnect lines.
copper, interconnect

corona charging an electric discharge in the air in the vicinity of semiconductor surface; air is ionized and ions are deposited on the surface; used in non-contact electrical characterization of semiconductors.
non-contact electrical characterization

cost of ownership, COO cost of implementing given process technology including cost of equipment, infrastructure, maintenance, materials, impact on existing tool set, etc.; typically determined for specific tool or set of tools needed to carry out specific processing goal.

coulomb unit of electric charge in SI system, unit symbol C; C = 1 Ampere x 1 sec. (charge transported by constant current of one ampere in one second).

Coulomb blockade, CB an effect causing increased resistance of semiconductor device in the vicinity of the tunnel junction at the very low bias voltage.
tunnel junction

Coulomb force the force between electrically charged particles.

Coulomb's law a quantitative expression defining electric force (Coulomb force) between two charged objects.

Coulomb scattering electrostatic interactions between electrons moving in semiconductor in the presence of electric field and atoms in the lattice, dopant ions, and defects.
scattering

coulombic interactions an attraction or repulsion interactions between electrically charged particles.

covalent bond material is held together by sharing electrons originating from the adjacent (bonded) atoms; in the case of silicon (atomic number 14) each atom shares its four outer electrons with each of the four adjacent atoms leaving no free electrons as remaining 10 electrons are tightly bonded to the nuclei; materials featuring covalent bond: e.g. group IV semiconductors Si, Ge, and C.
ionic bond, metallic bond

CP-4 etch a solution used to chemically polish Si surface; consists of $HF:HNO_3:CH_3COOH$ typically in $3:5:3$ ratio.
silicon etch

CPGA Ceramic Pin Grid Array; type of IC package.
package, ceramic package

CPU Central Processing Unit; microprocessor, i.e. the part of the computer which executes and controls computational operations.
microprocessor

Cr, chrome a metal used as an opaque (nontransparent to UV) part of the masks in photolithography; due to high optical density and sharp edges assures better than photoemulsion resolution of the pattern transfer process; also Cr photomasks are much more durable (scratch resistant) than emulsion masks.
chrome mask, photoemulsion, photomask

critical dimension, CD dimensions of the smallest features (width of interconnect line, contacts, etc.) which can be formed during semiconductor device/circuit manufacture using a given set of process tools.
design rules

critical thickness, h_c see *pseudomorphic material.*

critical temperature temperature above which semiconductor loses its extrinsic properties (due to the excessive generation of electron-hole pairs) and displays properties of an intrinsic semiconductor; temperature above which *p-n* junction ceases to exist and devices/circuits built into semiconductor are annihilated; wider bandgap (E_g) semiconductors feature higher critical temperature, and hence, can operate at higher temperatures.
thermal generation, intrinsic-, extrinsic semiconductor, thermal conductivity

cross-talk undesired capacitive interactions between two adjacent metal lines in the multilevel metallization scheme of an integrated circuit; to allow device operation at high frequency cross-talk must be minimized by using interlayer insulator featuring as low dielectric constant (k) as possible.
interconnect, low-k dielectric, ILD

crosslinking process of establishing new bonds in polymeric chains as a result of illumination; basis of the negative resist technology in photolithography; renders negative resist initially soluble in the developer insoluble.
chain scission, negative resist

CRT Cathode Ray Tube; imaging device based on e-beam scanning of phosphors covered back of the screen to generate image; replaced entirely by LCD, LED and OLED displays.
LCD, LED display, OLED display, phosphor

cryogenic aerosol high velocity frozen particles of inert gas such as CO_2 or Ar; directed toward the wafer surface cryogenic aerosol is used to remove particles from the wafer surface.
particles, dry cleaning

cryogenic pump high-vacuum pump operating in the pressure range from about 10^{-3} torr to about 10^{-9} torr; removes gas molecules from vacuum by trapping them on cold surfaces; efficient, clean pumping; used in high-end vacuum equipment in semiconductor manufacturing.
turbomolecular pump

cryosol spray same as *cryogenic aerosol.*

crystal a solid featuring periodic spatial arrangement of atoms referred to as long-range order; among crystals the single-crystal materials feature long-range order throughout the entire piece of material while poly-crystalline materials feature long-range order only within limited grains; crystal (crystalline) semiconductors play dominant role in semiconductor device engineering.
single-crystal, polycrystalline material, amorphous

crystal defects imperfections in the highly ordered crystallographic structure of a crystalline solid; if present in high densities, defects rather than the inherent physical properties of a given semiconductor will determine its electronic characteristics; major types of defects that can occur in the crystalline solid include point defects, line defects, planar defects, and volume defects.
point, line, planar, volume defects

crystal growth term refers to the process in the course of which a single-crystal material is being formed.
Czochralski crystal growth, Float-Zone growth, Bridgman method, crystal pulling, epitaxy

crystal lattice periodic, symmetrical three-dimensional arrangement of atoms in the crystal solids.
lattice point

Crystal Originated Pits, COP structural defects formed on the surface of the single-crystal semiconductor; not related to any external interactions/process malfunctions.
single-crystal, pit

Crystal Originated Particles, COP particles on the surface of the single-crystal wafer which result from crystal restructuring during process rather than from the external sources, e.g. process ambient.
particle

crystal (crystallographic) planes the planes in the crystal which define its orientation; key electrical, thermal, and mechanical properties of semiconductors are different along different crystal planes; Miller indices are used to define crystal planes.
Miller indices, surface orientation

crystal pulling foundation of the Czochralski single-crystal growth technique (CZ); process in which single-crystal seed is slowly withdrawn from the melt and crystalline material condenses at the liquid-solid interface gradually forming a rod-shaped piece of single-crystal material. *Czochralski crystal growth*

crystal unit cell see *unite cell*.

cubic system one out of seven systems of unit cells in crystallography; the unit cell is in this case in the shape of the cube; includes simple cubic cell, body-centered cell (BCC), and face-centered cell (FCC); most of key semiconductors crystallize following FCC lattice. *Face-Centered Cubic; diamond lattice, zincblend lattice, Wurzite structure*

current crowding a non-uniform distribution of the density of current in semiconductor; may cause device malfunction by localized over-heating, voids formation, enhanced electromigration, etc. *electromigration*

current gain the ratio of the output current to the input current in the transistor. *transistor*

current-voltage, I-V, measurement procedure used to determine electrical characteristics of semiconductor test structures and devices by measuring current flowing across the device as a function of applied voltage; several important characteristics of semiconductor material/ device can be derived from I-V measurements. *capacitance-voltage measurements*

custom integrated circuit see *ASIC*.

cut-off frequency the signal frequency at which transistor's current gain drops to unity. *current gain, transistor*

cut-off region an operating mode of the transistor in which its output current saturates at its lowest value; transistor is fully-"off". *saturation region*

C-V measurements see *capacitance-voltage measurements.*

CVD see *Chemical Vapor Deposition.*

CVS Constant-Voltage Stress, see *TDDB.*

Czochralski crystal growth, CZ process of obtaining single-crystal solids; the most common method used to fabricate large diameter silicon wafers (up to 450 mm in diameter); single-crystal material is pulled out of the melt in which single-crystal seed is immersed and then slowly withdrawn; desired conductivity type and doping level is accomplished by adding dopants to the melt; produces single-crystal with very low defect density, but with some oxygen and/or carbon contamination.
Bridgman growth, float-zone crystal growth, magnetic CZ, single-crystal

CZTS copper-zinc-tin-sulfur (or selenium) thin-film semiconductor materials used to fabricate solar cells with efficiencies around 10%.
conversion efficiency, solar cell

D

DAC Digital-to-Analog Converter.

damage process induced disruption of the chemical composition and/or structural/mechanical integrity of the wafer or materials on its surface; some damage can be reversed by thermal annealing (e.g. radiation damage), some is irreversible (e.g. mechanical damage) and can be removed by etching away damaged part of the material.
mechanical damage, radiation damage, RIE damage, RTA

damascene process process in which interconnect metal lines (e.g. copper) are delineated in dielectrics isolating them from each other not by means of photolithography and etching, but by means of the chemical-mechanical planarization (CMP); in this process: *(i)* interconnect pattern is first photolithographically defined in the layer of dielectric, *(ii)* metal is deposited to fill resulting trenches, and *(iii)* excess metal is removed by means of chemical-mechanical planarization; dual damascene is a modified version of the process.
chemical-mechanical planarization, dual damascene, ILD

dangling bond unsaturated (broken) interatomic bond at the semiconductor surface or in its bulk; electronically active; can interact with charge carriers in semiconductor.
surface state, interface trap

DARC™ Dielectric Anti-Reflective Coating; a trademark of Brewer Science, Inc.
anti-reflective coating

DBR Distributed Bragg Reflector; a multilayer structure implemented in surface emitting laser to produce high quality mirrors.
Surface Emitting Laser

DCFL Direct-Coupled FET Logic.
Field Effect Transistor

DDE Double Diffused Epitaxy; see *epitaxy*.

Deal-Grove model a model describing kinetics of thermal oxidation of silicon based on chemical reaction between silicon and oxidizing species; accurate for oxides thicker than about 30 nm; of limited use for oxide thinner than about 10 nm as it does not consider electrical interactions between oxidizing species and charge carriers in oxidized semiconductor; assumes surface reaction controlled oxide growth in early stage of oxidation (linear regime) and controlled by diffusion of oxidizing species through the oxide during extended oxidation (parabolic regime).
oxidation constants, thermal oxidation

Debye length a distance in semiconductor over which local electric field affects distribution of free charge carriers; decreases with increasing concentration of free charge carriers.
free charge carrier

deep depletion a non-equilibrium condition in the sub-surface region of semiconductor in MOS devices in which depletion layer is wider than in thermal equilibrium.
depletion

deep etching etching of geometrical features during fabrication of MEMS devices; quite commonly reaches tens of micrometers into the

bulk of the substrate wafer; Deep Reactive Ion Etching is a preferred deep etching method.
etch, MEMS, Deep Reactive Ion Etching

deep level the energy level "deep" in the energy gap (bandgap) of semiconductor, i.e. away from the edges of the conduction and valance bands and close to the intrinsic Fermi level; most commonly introduced by the metallic contaminants penetrating semiconductor.
energy level, DLTS, intrinsic Fermi level

Deep Level Transient Spectroscopy, DLTS spectroscopic method based on the electrical characterization of semiconductor structures allowing detection and quantitative analysis of the "deep" energy levels in the energy gap of semiconductor.
deep level

Deep Reactive Ion Etching, DRIE a method used to delineate deep geometrical features in silicon; needed to shape MEMS structures; based on the Reactive Ion Etching adapted specifically to the needs of the prolonged etching processes.
deep etching, Reactive Ion Etching, Bosch process

deep UV, DUV term refers to the very short wavelength UV (Ultraviolet) light (from 193 nm to about 300 nm); typically associated with 248 nm wavelength generated by KrF excimer laser, and 193 nm wavelength generated by ArF excimer laser.
excimer laser

deep UV photolithography a photolithography mode using deep UV for the photoresist exposure.
extreme UV photolithography

deep UV photoresist a photoresist used in deep UV photolithography, i.e. the photoresist sensitive to the UV light in the deep UV range.
deep UV, deep UV photolithography

defect an imperfection in terms of the structural integrity or chemical composition within the material system comprising semiconductor device; defects result in the deterioration of the device performance by interfering with charge transport, promoting breakdown, etc.
crystal defects

defect "decoration" a precipitation of the certain types of contaminants (mostly metallic) present in the semiconductor crystal around structural defects and creation of the electrically active centers interfering with the minority carrier lifetime.
minority carrier lifetime

degenerate semiconductor semiconductor that is so heavily doped that its Fermi level is located closer to one of the band edges (either conduction or valance) than *2 kT/q* or inside either conduction or valance band; acts more like a metal than semiconductor; some semiconductors are intrinsically degenerate.
energy bands, Fermi level, non-degenerate semiconductor

deionized water, DI water an ultra-pure water used in semiconductor manufacturing; produced by removing ions of dissolved minerals using reverse osmosis and ion-exchange processes; DI water should also be free from particles, bacteria, organics, and dissolved oxygen; purity of deionized water is determined based on its resistivity; target resistivity is 18 MΩ-cm.
rinsing, ozonated water

delta-doping formation of the stack of the individually doped *nm*-thick layers by means of the Molecular Beam Epitaxy (MBE) or MOCVD; resulting dopant distribution in the multilayer structure follows the Dirac delta function and is engineered to achieve specific device functions.
MBE, MOCVD

density of states the density of quantum states electron in semiconductor can occupy per unit energy per unit volume ($1/eV\text{-}cm^3$); an important physical concept which determines key physical properties of semiconductor such as charge transport, dependence of conductivity on temperature, optical absorption and others.

density of states function the density of quantum states electron can occupy as a function of energy.
density of states

denuded zone, DZ a very thin region in the semiconductor substrate wafer immediately adjacent to its surface from which contaminants

and/or defects were relocated toward the bulk of the wafer by means of gettering; devices are built into denuded zone.
gettering extrinsic, gettering intrinsic

Depleted Substrate Transistor, DST represents a class of MOSFETs built into the SOI substrate in such way that its active Si layer between source and drain is fully depleted creating a high conductivity channel; elimination of the junction capacitances results in the device featuring superior characteristics of the DST; same as "fully depleted SOI".
SOI, Fully-Depleted SOI

depletion condition within the semiconductor material system where concentration of majority carriers is lower than dopant concentration; under the normal thermal equilibrium condition concentration of majority carriers is the same as the concentration of dopant atoms.
thermal equilibrium, accumulation

depletion approximation an approximation which assumes that the edges of the depletion region are well defined and the transition to the adjacent regions is abrupt.
depletion region

depletion mode MOSFET a normally "on" MOSFET; i.e. device in which gate voltage V_G must be applied to turn it off; in other words MOSFET in which channel exists even at $V_G = 0$.
enhancement mode MOSFET, MOSFET

depletion region a region in semiconductor materials system in which the state of depletion is created; associated with the potential barrier present at the *p-n* junction or at the oxide-semiconductor interface (MOS devices) or at the metal-semiconductor contact.
depletion, space-charge region

depletion width a width of the depletion region.
depletion region

depth of focus, DOF distance along the optical axis over which features of the illuminated surface are in focus; one of key factors determining resolution of photolithography.
photolithography, resolution

depth profiling materials characterization method allowing determination of the distribution of elements in the solid in the direction normal to its surface; common in conjunction with SIMS, Auger Electron Spectroscopy and other methods of surface analysis; depth profiling is accomplished by simultaneous gradual removal of the material using ion milling and analysis of its chemical composition; depth profiling is broadly used in semiconductor R&D.
Auger Electron Spectroscopy, SIMS, ion milling

design rule the minimum dimension of the devices and interconnects comprising an integrated circuit adopted during the circuit's design stage; determined by the capabilities of process technology available.
critical dimension

DESIRE Diffusion-Enhanced Silylated Resist.
Photoresist

desorption process of freeing of the volatile species adsorbed at the solid surface as a result of external interactions such as reduced ambient pressure, or increased temperature, or illumination.
adsorption

developer in the lithographic processes a liquid used to dissolve parts of the resist following its selective exposure to UV light, or electron beam, or X-rays.
lithography

development process following resist exposure during which resist is locally dissolved in the developer.
developer

DFT Design For Testability; the circuit is designed such that it can performed the self-testing functions.

DHBT Double Heterojunction Bipolar Transistor.
bipolar transistor, HBT

DHET Double Heterojunction Field Effect Transistor.
Field Effect Transistor, HFET

DHF dilute HF; SiO_2 etching solution of 49% HF in water; typical composition: 1 part HF : 100 parts H_2O.
hydrofluoric acid

DI water see *deionized water*.

diamond a single-crystal carbon; wide bandgap semiconductor ($E_g = 5.4$ eV) featuring mobility of electrons and holes (300 K) of 1900 cm^2/Vsec and 1500 cm^2/Vsec respectively; excellent thermal conductivity; in theory a material featuring excellent semiconductor properties; in practice difficult to form in the shape and crystal quality compatible with commercial manufacturing of semiconductor devices; also, restrictions on the *p-n* junction formation due to the lack of adequate *n*-type dopants hamper progress in diamond semiconductor devices; thin-film diamond deposition by CVD was demonstrated.
carbon, wide-bandgap semiconductor

diamond lattice crystal structure which belongs to the face-centered cubic (FCC) class of crystals; several semiconductors, most notably silicon, crystallize following diamond lattice.
face-centered cubic, zincblend lattice

DIBL see *Drain-Induced Barrier Lowering*.

diborane, B_2H_6 gaseous compound commonly used in silicon manufacturing as a source of boron (B) which is *p*-type dopant in silicon.
diffusion source, dopant

dicing process of cutting semiconductor wafer into individual chips following completion of device fabrication (both discrete and integrated); in the case of large diameter wafer dicing is carried out by partially cutting the wafer along preferred crystallographic planes using high precision ultra-thin diamond blade saw; by wafer scribing in the case of smaller diameter wafers.
chip, die, scribing

die a single piece of semiconductor containing entire integrated circuit which has not yet been packaged; at the end of the manufacturing process wafers are separated into dies; synonymous with the term chip.
package, packaging, chip

die attachment process of attaching die (chip) to the package.
assembly, package

die separation see *dicing.*

dielectric a solid displaying insulating properties (energy gap typically wider than about 5 eV); its upper most energy band is completely empty, hence, dielectric features extremely low conductivity; by definition, characteristics of a dielectric are independent of the applied voltage; common dielectrics in semiconductor technology: SiO_2, SiN_4, Al_2O_3, etc.
insulator, high-k dielectric, low-k dielectric

dielectric relaxation time a time needed by a semiconductor to return to electrical neutrality after carrier injection or extraction.
carrier extraction, carrier injection

dielectric strength maximum electric field given material can withstand before breaking down; needs to be defined for the chemically pure and structurally homogenous material of a given thickness; thin-films tend to feature higher dielectric strength than their bulk counterparts.
breakdown, dielectric

diffusant a specie moving in the solid by diffusion.
diffusion, diffusion coefficient

diffused junction formation of the *p-n* junction by diffusion of dopants into semiconductor.
alloyed junction, implanted junction

diffusion in general, motion of species in solids in the direction of the concentration gradient; *(i)* diffusion of free charge carriers in semi-conductor; *(ii)* diffusion of dopant atoms in semiconductor at elevated temperature; also, term "diffusion" is commonly used in reference to the high temperature process during which dopant atoms are introduced into semiconductor by diffusion.
diffusion coefficient, diffusion current, doping

diffusion capacitance the capacitance of the forward biased *p-n* junction resulting from the delayed response of the minority carriers

concentration to the changes in bias voltage; minority carrier storage effect.
minority carrier storage

diffusion coefficient, *D*, for diffusion current unit cm^2/sec; determines rate with which charge carries (electrons and holes) move across semiconductor by diffusion at the lack of electric field; proportional to carrier mobility μ; higher for electrons than for holes because of the smaller effective mass of an electron.
Effective mass, Einstein relation, mobility

diffusion coefficient, *D*, for thermal diffusion unit cm^2/sec; determines rate with which elements move in a given solid by diffusion; depends on temperature; varies between elements by orders of magnitude, e.g. in the case of silicon at 1000 °C, *D* for gold (Au) is in the range of 10^{-7} cm^2/sec (fast diffusant) while for antimony (Sb) is in the range of 10^{-15} cm^2/sec (slow diffusant).
diffusant

diffusion current motion of charge carriers in semiconductor by diffusion, i.e. from the region of high carrier concentration to the region of low carrier concentration; diffusion current occurs in semiconductors in the absence of an electric field; in the presence of an electric field the drift current dominates electrical conduction in semiconductors.
drift current

diffusion, doping see *doping by diffusion.*

diffusion length term applies to the electron transport by diffusion; the distance over which concentration of free charge carriers injected into semiconductor falls to $1/e$ of its original value.
diffusion

diffusion pump a high-vacuum pump (from 10^{-3} torr to 10^{-7} torr) featuring relatively high pumping speed; in the past the most common high-vacuum pump in semiconductor processing; with time, mostly eliminated from the mainstream applications because of the oil vapor back-streaming into the vacuum system.
roughing pump, cryogenic pump, turbomolecular pump

diffusion source a source of dopant atoms used in semiconductor doping processes; gas, liquid, and solid sources are available for most of the common dopants.
diborane, phosphine, phosphorus tetrachloride

digital device, digital integrated circuit devices and circuits designed to realize digital (computational) functions; in digital systems output signal is either "on" or "off" responding to the input signal; designed to assure as rapid as possible switch of output signal from "on" to "off" state (and vice versa) with as little energy as possible.
analog device

digital ICs integrated circuits designed to carry logic and memory functions; technology driving class of integrated circuits.
digital device, logic IC, memory IC, analog ICs

Digital Light Processing®, DLP a display/projection technology developed by Texas Instruments; based on the arrays of MEMS microscopic mirrors integrated on the chip known as a Digital Micromirror Device (DMD).
MEMS, display

DIMOS Double Implanted MOSFET; configuration of the discrete power MOSFET in which the source and channel regions are formed using a double implantation process.
power device, DMOS

diode in general, a two-terminal element displaying asymmetric (rectifying) current-voltage characteristics; large current flows under the forward bias, very small current flows under the reverse bias; realized either as a vacuum diode or semiconductor diode.
semiconductor diode

DIP see *Dual Inline Package.*

direct bandgap the energy gap (bandgap) in semiconductor is configured such that the bottom of the conduction band and the top of the valence band coincide at the momentum $k = 0$.
indirect bandgap

direct bandgap semiconductor semiconductor featuring direct band-gap; in the case of a direct bandgap semiconductor the energy released during band-to-band electron recombination with a hole is predominantly in the form of the electromagnetic radiation (radiant recombination) without generation of phonons; wavelength of emitted radiation is determined by the energy gap of semiconductor; examples GaAs, InP, etc., only direct bandgap semiconductors can be used to fabricate light emitting devices (LEDs, lasers).
direct bandgap, indirect bandgap semiconductor, photon, phonon

direct plasma plasma process in which wafer is directly exposed to plasma and its products.
remote plasma

direct recombination electron in the conduction band releases energy and recombines directly with hole in the valence band, i.e. without involving any energy levels possibly located in the bandgap.
indirect recombination

direct tunneling electron tunneling directly across the entire width of the potential barrier rather than across the partial barrier; occurs in the MOS structures with ultra-thin oxide (< 3 nm) in which case electrons from the conduction band in semiconductor are tunneling across the oxide directly (i.e. without changing energy) into the conduction band of metal; probability of direct tunneling is a very strong function of the width of the barrier through which electrons are tunneling (oxide thickness in the case of MOS devices).
Fowler-Nordheim tunneling, tunneling

direct write lithography a lithography in which desired pattern is "written" in the resist directly using focused beam of electrons (e-beam lithography) or ions (ion-beam lithography); as opposed to masked lithography using masks to define the pattern.
masked lithography, e-beam lithography, ion-beam lithography

discrete device single semiconductor device such as transistor or diode mounted in the individual package; opposite to an "integrated" device such as transistor acting as an element of an integrated circuit.
integrated device

dislocation a two-dimensional (line) defect in single-crystal material; shift of the crystallographic planes with respect to each other; "edge" or "screw" dislocations are distinguished based on the direction of the shift.
crystal defects, line defect, edge dislocation, screw dislocation

dispersive medium a medium refractive index of which depends on the wavelength of incident light; semiconductor materials are dispersive.
refractive index

display a device converting electric signal into an image; the most efficient link between the digital domain and our senses; among the most common and the most important devices in our daily lives; all state-of-the-art displays use semiconductor devices as either light emitting or current controlling elements.
LED display, LCD LED display, organic display

distribution function (d.f.) term refers to the probability distribution function which in the case of semiconductors describes distribution of the energy states that can be occupied by an electron.
Bose-Einstein d.f, Fermi-Dirac d.f., Maxwell-Boltzman d.f

DLP see *Digital Light Processing.*

DLTS see *Deep Level Transient Spectroscopy.*

DMD, Digital Micromirror Device see *DLP*.

D-MOSFET, DMOS Double - Diffused MOSFET; configuration of the discrete power MOSFET in which the source and the channel regions are formed using a double-diffusion process.
power device, DIMOS

donor *n*-type dopant; element introduced to semiconductor to generate free electrons (by "donating" electron to semiconductor); must have one more valance electron than the host semiconductor; common donors in Si are P, As, Sb while in GaAs are S, Se, Sn, Si, and C.
acceptor, dopant

dopant an element introduced into semiconductor to *(i)* establish either *p*-type (acceptors) or *n*-type (donors) conductivity and *(ii)* establish

desired level of conductivity; common dopants in silicon: p-type, boron, B; n-type phosphorous, P, arsenic, As, and antimony, Sb.
acceptor, donor

dopant activation (ionization) process of dopant ionization causing release of the free electron (or free hole) after dopant has been incorporated into the semiconductor structure; requires a supply of thermal energy in excess of ionization energy.
ionization energy

dopant deactivation dopant atom may become inactive, i.e. is unable to release free charge carrier, if bonded with alien element in semiconductor; most common example in silicon: deactivation of p-type dopant boron, B, by hydrogen.
dopant, dopant activation

dopant redistribution term typically refers to the possible undesired, or induced on purpose, changes in the dopant profile during the high-temperature treatments to which semiconductor wafer may be subjected; driven by the dopant diffusion in the direction of the concentration gradient; in the case dopant redistribution is not desired only low thermal budget thermal treatments should be employed.
limited-source diffusion, RTA, thermal budget

dopants in silicon see *dopant*.

doping process of dopant introduction into semiconductor for the purpose of altering its electrical characteristics; allows control of the resistivity/conductivity of semiconductor by several orders of magnitude; also used to convert p-type material into n-type material and vice versa.
diffusion, ion implantation

doping by diffusion process of semiconductor doping carried out by means of thermal diffusion.
diffusion

doping by ion implantation process of semiconductor doping carried out by means of ion implantation.
ion implantation

doping during material growth establishing of the desired dopant concentration/conductivity type of semiconductor material by introducing dopant atoms during its formation process.
Czochralski crystal growth, epitaxy

dose typically understood as implantation dose; number of ions crossing one cm^2 of the surface of the implanted solid; unit ions/cm^2; controlled by ion beam current and implantation time; higher the dose, higher the concentration of implanted ions.
ion implantation, implantation dose

double (dual) gate an approach to the implementation of the MOSFET in which gate structure (gate oxide and gate contact) is formed on two sides of the channel, hence, "double"; doubles capacitance of the MOS gate without increasing device area; best realized in the MOSFETs featuring vertical channel such as FinFET.
FinFET, MOS, MOSFET, surround gate, gate all around

Double-Gate Transistor see *double gate*.

double patterning see *multiple patterning*.

double-side polishing a process in which both surfaces of the wafer, front and back, are polished instead of the standard one-side polishing; in some processing steps it is essential that the front and the back surface of the wafer feature the same optical (reflection, absorption) characteristics to assure the same temperature readings from both surfaces.
CMP

downstream plasma see *remote plasma*.

DPSSL Diode-Pumped Solid State Laser.
laser

DQN photoresist a two component photoresist: diazoquinone (DQ – photosensitive component) and novolac (N-resin); compatible with *g*-line (236 nm) and *i*-line (365 nm) exposure tools.
g-line lithography, i-line lithography

drain one of the three terminals in the Field Effect Transistor; heavily doped region located at one end of the channel in FETs; current is flowing out of the transistor through the drain.
Field-Effect Transistor, gate, source, drain engineering

drain current current flowing between source and drain regions of the FET; an output current of the FET.
source, Field Effect Transistor

drain engineering modifications of the drain and channel region adjacent to the drain in integrated MOS and CMOS devices; the goal is to improve characteristics of the MOS transistors with ultra-short channels.
drain, raised source-drain, short channel effects

drain extension a low doping extensions of the source and drain regions toward the channel; implemented by ion implantation; used in advanced CMOS to prevent premature breakdown of the drain-substrate junction.
drain engineering, lightly doped drain

Drain-Induced Barrier Lowering, DIBL parasitic short-channel effect in MOSFETs; results in the reduced control of the gate voltage V_G over the transistor's current as well as in the fluctuations of the threshold voltage V_T.
short-channel effects, threshold voltage

DRAM see *Dynamic Random Access Memory.*

DRIE see *Deep Reactive Ion Etching.*

drift motion of charge carriers in semiconductor under the influence of an electric field

drift current current (motion of charge carriers) in semiconductor caused by the electric field.
diffusion current

drift mobility a proportionality factor between drift velocity of charge carriers in semiconductor and the electric field.
mobility, drift velocity

drift velocity velocity of charge carriers moving in semiconductor under the influence of the electric field; as opposed to carriers in free space, carriers in semiconductor are not "infinitely" accelerated by the electric field because of the scattering, and thus, they reach a finite velocity regardless of the period of time over which the field is acting; at a given electric field drift velocity is determined by the mobility of charge carrier.
scattering, mobility

drive current term used in reference to the output (drain) current of the MOSFET operating as an element of the digital IC; an output current of the MOSFET/CMOS in "on" state.
digital ICs, drain current

drive-in a high temperature (typically > 800 °C) operation performed on the semiconductor wafer in an inert ambient; causes motion of dopant atoms in semiconductor in the direction of the concentration gradient (diffusion); used to drive dopant atoms deeper into semiconductor; typically a second step in semiconductor doping by diffusion designed to establish desired dopant distribution and junction depth.
diffusion, limited-source diffusion, redistribution, junction depth

dry cleaning a process of contaminants removal from the wafer surface carried out in the gas-phase; driven by either conversion of contaminant into volatile compound through chemical reaction, or its "knocking" off the surface via momentum transfer (e.g. cryogenic aerosol), or through the lift-off during slight etching of contaminated surface.
wet cleaning, surface conditioning, cryogenic aerosol

dry etching an etching process carried out in the gas-phase; can be either purely chemical (plasma etching), purely physical (ion milling) or combination of both (Reactive Ion Etching, RIE).
wet etching, plasma etching, ion milling, RIE

dry oxidation process of thermal oxidation of silicon carried out in moisture-free (dry) oxygen; slower than wet oxidation, but results in SiO_2 featuring superior electrical integrity.
wet oxidation, thermal oxidation

dry oxide a thin layer of silicon dioxide (SiO_2) grown on the Si surface by dry oxidation; gate oxides in silicon MOSFETs are dry oxides as they feature superior electrical integrity.
dry oxidation, gate oxide, wet oxide

dry pump an entirely dry roughing (down to the vacuum of about 10^{-3} torr) pump; common in advanced semiconductor process equipment because it is oil-free, and hence, "clean"; e.g. Roots pump.
roughing pump, Roots pump

drying a process of removing water/moisture from the wafer surface and rendering it dry at the molecular level.
IPA drying, Marangoni drying, spin drying

DST see *Depleted Substrate Transistor.*

DSW Direct Step on Wafer; the term is referring to the advanced resist exposure technique used in photolithography.
step-and-repeat projection

DTMOSFET Dynamic Threshold MOSFET.

dual damascene a modified version of the damascene process employed to define metal interconnect geometry in an IC using CMP process instead of metal etching; in dual damascene two interlayer dielectric patterning steps and one CMP step create a pattern which would require two patterning steps and two metal CMP steps when using conventional damascene process.
Chemical-Mechanical Planarization, damascene

Dual Inline Package, DIP common type of package used to mount and enclose semiconductor chips; input/output pins are located in-line along two sides of the package; comes in a variety of material configurations.
package

DUV see *deep UV.*

DUV photolithography see *deep UV photolithography.*

DUV photoresist see *deep UV photoresist.*

DWCNT   Double Walled Carbon Nanotube.
single walled nanotube, carbon nanotube

Dynamic Random Access Memory, DRAM   memory cell in which digital information (data) is stored in the volatile state; a key component of the digital circuits; specific feature of DRAM cell is that besides transistor each cell requires capacitor(s) to store information; capacitors must be recharged (refreshed) periodically, hence, "dynamic".
memory cell, SRAM

DZ   see *denuded zone*.

E

Early effect   reduction of the width of the base in bipolar transistor due to the widening of the base-collector junction with increasing base-collector voltage; Early effect may be observed on the output charac-teristics (collector current *vs.* base-collector voltage) as a slight increase of the collector current in the saturation region.
base, bipolar transistor

E_{bd} breakdown field; electric field in the MOS gate oxide at which oxide breaks down, i.e. looses its insulating properties in an irreversible manner; E_{bd} is measured by increasing voltage (ramp voltage) applied to MOS gate while monitoring current flowing across the gate oxide.
oxide breakdown, ramp voltage, charge-to-breakdown

EBDW   see *Electron-Beam Direct Write*.

Ebers-Moll model   the basic model (equivalent circuit) describing op-eration of bipolar transistor primarily in switching applications.
hybrid-pi model

EBIC   see *Electron Beam Induced Current*.

EBL   see *electron beam (e-beam) lithography*.

EBL column   part of the electron-beam lithography (EBL) apparatus that forms and controls electron beam used to exposed resist; a multi-component, very complex system.
electron beam (e-beam) lithography

ECL Emitter Coupled Logic; based on the bipolar transistor technology; very fast switching gate but "bulky" and power consuming.
I^2L

ECR Electron Cyclotron Resonance; used to generate high density plasma; efficiency of ionization of gas (density of plasma) is increased by the magnetic field which is forcing electrons into cyclotron resonance.
ECR plasma, high density plasma

ECR plasma high density plasma generated using Electron Cyclotron Resonance; used primarily in etching applications; ECR plasma allows efficient etching with limited surface damage; typically implemented in the "remote plasma" configuration.
helicon plasma, ICP, remote plasma

edge dislocation a line defect; a slip of the part of crystal with dislocation line moving parallel to stress in the lattice.
line defect, screw dislocation

Edge Emitting Laser, EEL a structure of the semiconductor laser; emitted light comes out from the edge of the *p-n* junction in the direction parallel to the junction plane.
Surface Emitting Laser

Edge Emitting Light Emitting Diode, EELED a structure of the semiconductor Light Emitting Diode; emitted light comes out from the edge of the *p-n*-junction in the direction parallel to the junction plane.
Surface Emitting LED

EEL see *Edge Emitting Laser*.

EELED see *Edge Emitting LED*.

EELS Electron Energy Loss Spectroscopy.

EEPROM Electrically Erasable Programmable Read Only Memory; a nonvolatile memory; also known as *E^2PROM*; a memory cell in which information (data) can be erased and replaced with new information (data) with the application of an electric pulse.
nonvolatile memory

effective mass, m^* an effective mass of electrons and holes in semiconductor is defined taking into account forces exerted on the carriers by the atoms in the given crystal; different in different semiconductors, and hence, mobility of the charge carriers is different in different semiconductors.
mobility

efficiency term commonly applies to the energy conversion efficiency of the solar cells; defined as the percentage of the incoming energy of sun light converted into the output electrical energy; varies from about 1% to over 45% depending on the materials used and structure of the solar cell.
solar cells, external quantum efficiency, internal quantum efficiency

EFG, Edge-Defined Film-Fed Growth crystal growth method used, for instance, to form sapphire sheets and ribbons.
sapphire

E_g a symbol of energy gap (bandgap) in solids; see *energy gap*.

Einstein relation a relation between diffusion coefficient and mobility of charge carriers in semiconductor; diffusion coefficient D is proportional to mobility with kT/q as a proportionality factor.
diffusion coefficient, mobility

elastic collision an ideal elastic collision is the one during which there is no loss of kinetic energy and momentum is conserved; with an exception of the special cases of scattering, collisions between species in solids are never perfectly elastic.
inelastic collisions

electroluminescence emission of light by the material stimulated by the current flowing though it or an electric field; allows direct conversion of electric energy into visible light without generation of heat.
photoluminescence, luminescence

electromigration deleterious effect plaguing some metals, but in particular aluminum; physical motion of atoms away from the areas where current density is very high; caused primarily by frictional force between metal ions and flowing electrons; results in the break in the metal line and formation of hillocks; common cause of the malfunction

of the aluminum interconnect network in integrated circuits; main reason for which Al interconnects were replaced with copper interconnects in advanced IC technology.
aluminum, copper, interconnect line, hillock

electron a negatively charged particle in an atom; carrier of the smallest (elemental) electric charge of 1.6×10^{-19} C; can be freed from its binding to the nuclei and become a mobile carrier of the negative charge in semiconductors; responsible for electrical conductivity of solids as well formation of inter-atomic bonds in conducting solids; serves as an "information carrier" in electronic devices and circuits; in the presence of very high electric field in ultra-small semiconductor structures exhibits properties of the wave.
electronic device, photon, hole, elemental charge, wave-particle duality

electron affinity for semiconductors the term is defined as a difference between the vacuum level and the bottom of the conduction band in an atom.
conduction band, vacuum level

electron beam (e-beam) geometrically confined and accelerated by electromagnetic forces stream of electrons directed toward desired target; carries kinetic energy that depends on acceleration; can be focused down to 1 nm in diameter; direction of the beam can be rapidly altered in the controlled fashion to scan the target; the shape of the beam (cross section) can be precisely controlled and also rapidly changed; allows delivery of the significant amount of energy to the bombarded solid over the highly confined area; e-beam finds broad range of uses in semiconductor science and engineering.
electron gun

electron beam direct write term applies primarily to the e-beam lithography where a computer driven e-beam writes desired pattern directly into the layer of resist; a mask-less pattern definition technique which is also extensively used in making masks for photolithography.
electron beam lithography, mask making, pattern generation

electron beam (e-beam) evaporation a thin-film deposition technique; material is evaporated as a result of highly localized heating/melting resulting from the bombardment with high energy e-beam; evaporated material is very pure; bombardment of metal with electrons is

accompanied by the generation of low-intensity X-rays which may create defects in the oxide present on the surface of the substrate; typically, an anneal is needed to eliminate such defects.
evaporation, filament evaporation, sputtering, PVD, thermal evaporation

electron beam (e-beam) heating generation of heat in the material bombarded with high energy electrons as a result of the momentum transfer.
Electron beam evaporation, Rapid Thermal Processing

Electron Beam Induced Current, EBIC part of the SEM and STEM material characterization methods; based on the e-beam stimulated generation of electron-hole pairs (current) in the bombarded semiconductor.
electron beam, SEM, STEM

electron beam (e-beam) lithography, EBL a lithography technique which uses focused beam of electrons to expose the resist; no mask is used as pattern is "written" directly into the resist by rapidly scanned electron beam; very high pattern transfer resolution below 10 nm is possible; resolution is limited by the proximity effect; EBL is commonly used to manufacture high resolution masks for photolithography.
lithography, raster scan, variable shape beam, vector scan, proximity effect

electron beam (e-beam) resist a resist used in e-beam lithography; formulated in such way that the desired chemical reactions are promoted in the layer of resist (e-beam resist) specifically by the impinging high energy electrons and not by UV light.
resist

Electron Beam (e-beam) Vapor Deposition see *electron beam evaporation.*

electron effective mass see *effective mass.*

electron gas, 2-dimensional see *Two-Dimensional Electron Gas.*

electron gun a device installed in vacuum systems to produce electrons; most commonly electrons are emitted from the solid hot filament then

focused into the beam and accelerated by the magnetic field toward the target.
electron beam, field emission

electron-hole pair whenever electron acquires energy sufficient to "move" from the valence band to the conduction band of the semiconductor, a free hole is created in the valence band; an electron-hole pair is generated (generation); an electron-hole pair is annihilated when electron and hole recombine (recombination).
generation, recombination

electron microscopy see *Scanning Electron Microscopy*.

electron mobility a proportionality factor between electron drift velocity and electric field as well as electron concentration and conductivity of semiconductor; unit $cm^2/Vsec$; measure of electron scattering in semiconductor; the same way as the effective mass of an electron, electron mobility is different for different semiconductors; electron mobility at 300 K for: Si - 1500 $cm^2/Vsec$., GaAs - 7500 $cm^2/Vsec$., GaN - 1000 $cm^2/Vsec$., Ge - 1900 $cm^2/Vsec$., 6H-SiC - 400 $cm^2/Vsec$.; higher electron mobility makes semiconductor better suited for high speed applications both analog (high-frequency communication) and digital (faster switching transistors).
mobility, effective mass, scattering, drift velocity

Electron Paramagnetic Resonance, EPR also referred to as *Electron Spin Resonance*; in common usage a name of the material characterization method detecting very fine structural defects in semiconductors and dielectrics; detects changes in the spin state of an electron in the presence of the very high magnetic field.
electron spin

Electron Projection Lithography, EPL modification of the e-beam lithography for the purpose of increasing throughput of the process by eliminating the time consuming scanning.
electron beam lithography, PREVAIL, SCALPEL

electron resist see *electron beam resist*.

electron scattering see *scattering*.

Electron Spectroscopy for Chemical Analysis, ESCA a surface characterization method based on the determination of energy of photoelectrons ejected from the solid surface region as a results of either UV illumination (UPS - UV Photoelectron Spectroscopy), or X-ray irradiation (XPS - X-ray Photoelectron Spectroscopy).
UPS, XPS

electron spin, spin electron spin is an inherent property of an electron; binary in nature (spin $s = 1/2$ or $s = -1/2$ only); this electron characteristics is a foundation for the emerging new generation of digital semiconductor devices (see *spitronics*) in which electron's spin rather than its electric charge controls device operation.
spintronics

Electron Spin Resonance see *Electron Paramagnetic Resonance.*

electronic device a semiconductor device operation of which is based on the interactions of electrons and holes (generation-recombination, transport) in the solid; electron is acting as information/energy carrier.
photonic device, semiconductor device

electronic grade poly Si 99.99999% pure polycrystalline silicon used as a starting material in the single-crystal growth process.
single-crystal growth

electronics technical domain concerned with devices based on electronic interactions.
electron, photonics, plasmonics, spintronics

electro-optic effect defines dependence of the optical properties of a material (specifically its refractive index) on the electric field; observed in several semiconductors, for instance GaAs; an important effect bridging electronics and photonics.
refractive index

electroplating a thin layer of metal is plated on the surface of the biased wafer immersed in the electrolyte containing metal ions; a layer of the conducting seed material must be deposited on the surface of an insulator prior to electroplating; a method used in IC manufacturing to deposit copper.
copper interconnect

elemental charge, q 1.6×10^{-19} C; electric charge of an electron.
electron

elemental semiconductors single element semiconductors; a device-grade elemental semiconductors are elements from the group IV of the periodic table: Si, Ge, and C; as opposed to compound semiconductors which do not appear in nature; silicon is the most common elemental semiconductor.
semiconductor, silicon, germanium, carbon, tin, compound semiconductors

elevated drain a drain region in the MOSFET/CMOS is extended above the surface of the wafer by means of selective epitaxy; needed in ultra-short channel MOSFETs where drain region must be extremely shallow (below 10 nm); elevation above the surface reduces drain's series resistance while maintaining its shallowness; the same concept as a raised drain.
drain, series resistance, raised source-drain, selective epitaxy

elevated source conceptually the same as *elevated drain.*
source, series resistance, raised source-drain

ellipsometry the most common way of measuring thickness of thin films; based on the detection of phase shift of the plane polarized incident light beam during beam reflection from the surface; the film must be transparent to the incident light for the ellipsometry to work.
spectroscopic ellipsometry

ELO see *Epitaxial Lateral Overgrowth.*

emissive display a display composed of the elements that directly convert electric signal into light, e.g. light emitting diodes.
LED display, non-emissive display

emitter a very high conductivity region in semiconductor devices acting as a source of free charge carriers which are injected into the adjacent region, e.g. into base in the bipolar transistor.
base, bipolar transistor

emitter injection efficiency factor a factor defining efficiency of majority carrier injection from the emitter to the base (where they

65

become minority carriers) of a bipolar transistor; should be as close to 1 as possible.
base transport factor

emitter push effect occurs in *n-p-n* junction bipolar transistors; part of the boron doped base region which is in contact with phosphorous doped emitter region is deeper than the part of the base which is not in contact with emitter; this effect is due to the dissociation of phosphorus-vacancy pairs which enhance diffusion of boron deeper into the substrate under the emitter region.
emitter, base, n-p-n transistor, phosphorous-vacancy pair

endpoint during etching operations point in time when etching of the material meant to be etched is completed and continuation of etching results in an overetching.
overetching

endpoint detection during the gas-phase etching a determination the endpoint; common endpoint detection methods include interferometry or spectral analysis of the etching plasma (optical emission spectroscopy).
endpoint, laser interferometry, optical emission spectroscopy

energy band(s) the ranges of energy in semiconductor which an electron may have (valance band and conduction band) or may not have (forbidden gap).
conduction band, valance band, energy gap

energy gap, forbidden band, bandgap, E_g the energy band separating conduction and valence bands in the solid; no electron energy levels are allowed in the forbidden band; there is no energy gap in the metals in which case conduction and valence bands overlap; solids featuring energy gap are defined as either semiconductors or insulators based on the width of the energy gap; values of E_g (at 300 K) for common semi-conductors: InSb - 0.17 eV, Ge - 0.67 eV, Si - 1.12 eV, GaAs - 1.43 eV, GaP - 2.26 eV, 6H-SiC - 2.9 eV, GaN - 3.5 eV, and insulators Ta_2O_5 - 4.2 eV, TiO_2 - 5 eV, Si_3N_4 - 5.1, Al_2O_3 ~ 5 eV, SiO_2 - 8.0 eV.
conduction band, direct bandgap, indirect bandgap, valence band

engineered wafer a bulk silicon wafer which is modified by burying a layer of an oxide underneath the surface or by formation of an epitaxial layer on its surface, or by gettering.
bulk wafer, SOI wafer, epitaxial extension, gettering

enhancement mode MOSFET a normally "off" MOSFET; i.e. device in which gate voltage must be applied to turn it on; in other words MOSFET in which channel does not exist, and hence, transistor does not conduct at the gate voltage $V_G = 0$.
depletion mode MOSFET

enhancement techniques term typically refers to the techniques employed to extend the use of photolithography into deep sub-100 nm regime while maintaining 193 nm exposure wavelength; e.g., phase-shift masks, immersion photolithography, multiple exposure, computational lithography, etc.
immersion lithography, phase-shift mask, computational lithography, multiple exposure

EOT see *equivalent oxide thickness.*

epi layer in common semiconductor terminology a short version of the term epitaxial layer.
epitaxial layer

epitaxial deposition see *epitaxy.*

epitaxial extension formation of an epitaxial layer on the surface of bulk wafer.
engineered wafer, bulk wafer, SOI wafer

epitaxial growth see *epitaxy.*

Epitaxial Lateral Overgrowth, ELO local epitaxial growth which initially occurs in the direction normal to the surface of the substrate, but then proceeds preferentially in the direction parallel to the surface of the substrate; vertical growth starts from the single-crystal seed area, but lateral growth continues over non-crystalline portion of the substrate; can be used to form SOI substrates.
epitaxy, SOI

epitaxial layer a layer grown as a result of epitaxial deposition, or epitaxy; its crystallographic structure reproduces structure of the substrate, but doping level and conductivity type of epitaxial layer is controlled independently of the substrate; can be made chemically purer than the substrate, but substrate surface defects are reproduces in the

epitaxial layer; silicon substrates with epitaxial layers are commonly used in CMOS and bipolar device technology.
epitaxy

epitaxy an "ordered deposition"; a process by which thin layer of single-crystal material is deposited on single-crystal substrate; epitaxial growth occurs in such way that the crystallographic structure of the substrate is reproduced in the growing material; also crystalline defects of the substrate surface are reproduced in the growing material; a key process in advanced semiconductor device engineering.
single-crystal

epitaxy by CVD epitaxial growth implemented by means of chemical vapor deposition, i.e. by chemical reaction in the gas-phase product of which is a solid to be epitaxially deposited on the exposed substrate; typically carried out at temperatures above 1000 °C; with extremely thorough surface preparation and adequate selection of reactants temperature of CVD epitaxy can be reduced to as low as 500 °C - 600 °C; low-temperature CVD epitaxy can also be accomplished with Metal-Organic CVD.
Metal-Organic CVD, Molecular Beam Epitaxy

epitaxy by MBE see *Molecular Beam Epitaxy*.

EPL see *Electron Projection Lithography*.

EPR see *Electron Paramagnetic Resonance*.

EPROM Erasable Programmable Read Only Memory; a nonvolatile memory; a memory cell in which information (data) can be erased and replaced with new information (data).
nonvolatile memory

equivalent oxide thickness, EOT a number [nm] used to compare performance of the high-k dielectric MOS gates with performance of SiO_2 based MOS gates; gives thickness of SiO_2 gate oxide needed to obtain the same gate capacitance as the one obtained with a thicker than SiO_2 dielectric featuring higher dielectric constant k; e.g. EOT of 1 nm would result from the use of 5 nm thick dielectric featuring $k = 19.5$ (k of SiO_2 is 3.9).
high-k dielectric, gate oxide

error function a function which in semiconductor physics describes distribution of the dopant atoms during the pre-deposition (unlimited source diffusion) process in the doping by diffusion.
pre-deposition, doping by diffusion

ESCA see *Electron Spectroscopy for Chemical Analysis.*

ESL Etch Stop Layer; used in various device fabrication schemes to prevent undesired etching; see *etch stop.*

ESR Electron Spin Resonance; see *Electron Paramagnetic Resonance.*

EST Emitter Switched Thyristor; see *thyristor.*

ET SOI see *Extremely-Thin SOI.*

etch, etching subtractive process in the course of which solid is either dissolved in liquid chemicals (wet etching) or converted into gaseous compound (dry etching); key processes in top-down semiconductor manufacturing scheme.
dry etching, wet etching, top-down process

etch anisotropy see *anisotropic etch.*

etch isotropy see *isotropic etch.*

etch mask a material blocking etching in selected areas on the wafer surface; typically photoresist is acting as an etch mask.
photoresist

etch rate the rate at which given material is being etched; typically expressed in μm/min.
etch

etch selectivity defines the difference in the etch rate of various materials; etch process is designed such that selected material is etched at the much higher rate than other materials on the wafer surface; expressed in terms of the ratio of the etch rates of two materials.
non-selective etching, selective etching, etch rate

etch stop a material featuring drastically slower etch rate than the material subjected to etching; a layer of "etch stop" material is placed underneath etched material to stop etching process.
endpoint, ESL

ESL Etch Stop Layer.
etch stop

EUV see *Extreme UV.*

EUVL see *Extreme UV Lithography.*

evaporation common technique used to deposit thin-film materials; material to be deposited is heated in vacuum (10^{-6} torr - 10^{-7} torr range) until it melts and starts evaporating; vapor of material is condensing on the cooler substrate exposed to the vapor; common technique in thin-film metal deposition; not suitable for high melting point materials; one of the PVD (Physical Vapor Deposition) methods.
electron beam evaporation, filament evaporation, PVD, sputtering, torr

excess carriers charge carriers in semiconductor that are in excess of their thermal equilibrium concentration; subject to the process of recombination driven by the need to reach equilibrium; kinetics of this process determines carrier's lifetime.
recombination, lifetime

excimer laser a chemical laser; a laser capable of generating very short wavelength (below 250 nm) UV light, i.e. shorter than achievable using conventional UV lamps; generates highly collimated beam featuring uniform distribution of power across its diameter; examples of excimer lasers: KrF (248 nm emission wavelength), ArF (193 nm); commonly used as a source of UV light in high resolution photolithography.
deep UV

exciton an electron and a hole at the distance from each other, but bound by coulombic interactions and moving as a pair; excitons define light absorption and emission characteristics of excitonic materials, e.g. organic semiconductors.
coulombic interactions, organic LED, organic solar cells

excitonic material material in which excitonic processes rather than photonic processes define light absorption and emission characteristics.
exciton

excitonic solar cells, XSC solar cells which operation is based on the excitonic interactions rather than on the photovoltaic effect involving light stimulated generation of electrons and holes; most organic solar cells fall into the category of XSCs.
exciton, organic solar cells, photovoltaic effect

excitonics a technical/scientific domain focusing on the development and applications of the excitonic materials and devices.
exciton, excitonic material

exposure illumination of the resist with light featuring specific wavelength (UV in the case of photolithography) for the purpose of stimulating desired photo-chemical reactions in the layer of resist.
chain scission, cross linking, resist, photoresist, stepper

exposure wavelength a wavelength of light used to expose photoresist in photolithography; defines classes of photolithographic techniques; in general, shorter the wavelength, smaller geometrical features can be created on the wafer surface, hence, higher the resolution of the photolithographic pattern transfer process.
DUV, EUV, i-line lithography, g-line lithography, resolution

external emission an emission of electron from the solid under the influence of high electric field, or emission of photon from semiconductor as a result of the electron-hole recombination.
internal emission

external quantum efficiency, light emitters the ratio of the number of photons emitted from semiconductor device to the number of photons generated in the device.
photon

external quantum efficiency, solar cells the ratio of the number of photons arriving at the surface of the solar cell to the number of photons generating free charge carriers in the solar cell, i.e. carriers contributing to the short-circuit current of the cell.
solar cell, short-circuit current, internal quantum efficiency

extreme UV, EUV extremely short wavelengths in the 11 - 14 nm range of the electromagnetic spectrum; also referred to as a "soft X-ray"; generated by very high energy plasma; very highly absorbed by any medium, hence, requires high vacuum for transmission; for the same reason EUV cannot be controlled using optical lenses; used for the resist exposure in Extreme UV Lithography.
deep UV, UV, extreme UV lithography

Extreme UV Lithography, EUVL a photolithography using extreme UV (typically 13.5 nm wavelength) for pattern definition; the EUVL offers the highest pattern transfer resolution achievable with photolithography; requires reflective optics (very high precision mirrors) and reflective masks as the conventional transmission lenses and masks are not feasible with EUV because of the very high absorption of light in 13.5 nm wavelength range; its introduction to the commercial IC manufacturing is challenged by the technical problems, very high cost, and continued improvements in DUV lithography.
extreme UV, reflective mask, DUV lithography

extreme UV photoresist a resist sensitive to the UV light featuring wavelength in the 11 - 14 nm range.
photoresist

Extremely-Thin SOI, ETSOI the SOI substrate wafer with an extremely thin (below 10 nm) Si active layer; also referred to as *Ultra-Thin Body SOI (UTB SOI)*.
SOI, active silicon layer

extrinsic gettering semiconductor wafer is exposed to external physical interactions designed to induce stress at the back surface of the wafer; during subsequent thermal treatment defects and/or contaminants will move preferentially toward the stressed region, and hence, away from the top surface of the wafer; "denuded zone" is formed at the top surface.
gettering, intrinsic gettering, metallic contaminants, denuded zone

extrinsic semiconductor doped semiconductor featuring either *p*- or *n*-type conductivity.
intrinsic semiconductor, p-type conductivity, n-type conductivity

extrinsic temperature range temperature range in which all dopants in semiconductor are ionized and electron concentration remains constant.
dopant activation

extrusion coating process in which a viscous liquid precursor (e.g. photoresist) is applied to the surface in a spiral motion by extrusion instead of being dispensed on the surface and distributed by a spin-on process.
Physical Liquid Deposition, spin-on deposition

F

Fabry-Perot cavity the most common cavity structure in semiconductor lasers; less complex to implement than other types of cavities.
laser

Face-Centered Cubic, FCC, cell the crystalline structure of key semiconductors is based on the FCC cell; FCC cell, along with body-centered cubic (BCC) and simple cubic cell, belongs to the cubic system; diamond lattice (e.g. Si, Ge) and zinc blend lattice (e.g. GaAs, GaP) represent FCC sublattices.
crystal lattice, cubic system, body-centered cubic

faceting distortion of the pattern on the surface of the wafer subjected to sputtering process; results from the fact that the sputtering yield depends on the angle of incidence of ions impinging on the surface.
sputtering, sputtering yield

FAMOS Floating Gate Avalanche MOS.

fast surface state surface state capable of fast exchange of charge (charging and discharging) with the bulk of semiconductor in response to the fast changes of the bias voltage.
surface state, slow surface state

FCC see *Face Centered Cubic.*

FD FinFET Fully-Depleted FinFET
FinFET, depleted region

FD MOSFET see *Fully-Depleted MOSFET.*

FD SOI see *Fully-Depleted SOI.*

FEOL see *Front-End-of-Line*.

Fermi-Dirac distribution function (d.f.) a formula describing probability of the state being occupied by an electron depending on the state's
energy level.
Bose-Einstein d.f, Maxwell-Boltzman d.f

Fermi energy the maximum energy of an electron in a metal at 0 K.

Fermi level the energy level in solids at which the Fermi-Dirac
distribution function is equal to 0.5; in other words the probability of an
electron to occupy this energy level is 0.5.
Fermi-Dirac distribution function

Fermi level pinning, FLP imperfections of various nature of the
semiconductor surfaces may prevent the surface potential to respond
to the changes of the voltage applied to the metal contact in metal-
semiconductor and metal-insulator-semiconductor structures; Fermi level
pinning is a manifestation of this condition; the M-S and MIS/MOS
devices are rendered dysfunctional when the Fermi level is pinned.
surface potential, MOS, metal-semiconductor contact

Fermi potential potential difference between Fermi level and the
intrinsic Fermi level in the bulk of semiconductor; varies with semiconductor doping level.
intrinsic Fermi level

ferroelectric crystal a material (typically oxide) featuring permanent
(i.e. even without external electric field) electric dipole moment, and
hence, displaying spontaneous polarization (centers of positive and
negative charges of the crystal do not coincide); this state can be altered
by applying voltage to the crystal; as opposed to dielectric crystals which
properties are not dependent on the electric field; memory effect possible
with ferroelectric crystals; used in FRAMs.
FRAM, dielectric

ferromagnetic semiconductors same as magnetic semiconductors;
materials displaying semiconducting and ferroelectric properties; allow
integration of logic and memory functions on the same chip; foundation

74

of spintronics; prime example: GaMnAs, i.e. GaAs in which small number (~ 5%) of Ga atoms is replaced with Mn atoms.
spintronics, magnetic semiconductor

FET   see *Field-Effect Transistor.*

FIB   Focused Ion Beam; geometrically confined stream of ions.

Fick's laws   describe diffusion in solids; 1st and 2nd Fick's law; 1st Fick's law describes motion by diffusion of an element in the solid in the direction of the concentration gradient; 2nd Fick's law determines changes of the concentration gradient in the course of diffusion (function of time and diffusion coefficient).
diffusion

field effect   a potential applied to the metal contact on semiconductor surface featuring high potential barrier (potential barrier prevents current flow between metal and semiconductor) or on the oxide covered semiconductor surface; causes changes in the density/distribution of electric charge in the near-surface region of semiconductor; in this way conductivity of the near-surface region of semiconductor can be controlled by the potential applied to the metal contact (gate); as the name indicates the operation of the Field Effect Transistors (FETs) relies on the field effect.
Field Effect Transistor, potential barrier, space-charge region

Field Effect Transistor, FET   a transistor which operation is based on the field effect; FETs output current (source-drain current) is controlled by the voltage applied to the gate which can be either an MOS structure (MOSFET), a *p-n* junction (JFET), or metal-semiconductor contact (MESFET); FET is an unipolar transistor, i.e. its current is controlled by majority carriers only; the class of transistors the most commonly employed in practice particularly in the MOSFET version.
JFET, MESFET, MOSFET

field emission, field electron emission   emission of an electron from the solid (typically into the vacuum) promoted solely by the electric field, i.e. with no thermal stimulation; also known as cold emission; implemented using cold cathodes.

field oxide, FOX   a relatively thick oxide (typically 100 - 500 nm thick) formed to passivate and protect semiconductor surface outside of the active device area; part of any semiconductor device; typically formed by wet oxidation (in the case of silicon) or CVD.
CVD, wet oxidation

filament evaporation   thermal evaporation; source material is attached to the filament (refractory metal, e.g. tungsten) and melted by high current flowing through the filament; alternative approach is to use a "boat" (crucible) made out of the refractory metal which contains material to be evaporated; current sufficient to melt source material is passed through the "boat".
evaporation, electron-beam evaporation

fill factor   parameter which is a measure of performance of the solar cell; expressed as the ratio of the output electric power over the product of short-circuit current and open-circuit voltage; should be as close to 1 as possible; typically in the ~ 0.7 - 0.9 range.
solar cell, open-circuit voltage, short circuit current

FinFET   a configuration of the MOSFET in which transistor channel is positioned vertically in the "fin"-like shape; the gate is wrapped around the "fin" (gate all-around) allowing increased gate area, hence, gate capacitance and improved electrostatics at the reduced area occupied by the transistor on the wafer surface; FinFET is a leading transistor architecture in 22 nm and below MOSFET/CMOS technology generations.
gate all-around

fixed charge   see *oxide fixed charge*.

Fixed Shape Beam, FSB   the FSB is an alternative to variable shape beam concept in e-beam lithography.
e-beam lithography, variable shape beam

flash memory   a nonvolatile semiconductor memory; an EPROM or EEPROM with ability to block erasure of stored information (data), but with ability to be electrically reprogrammed.
EPROM, EEPROM

flat-band voltage, V_{FB} the key concept in MOS devices; a gate voltage at which no electric charge is present in semiconductor and therefore, no voltage drops across it; in the MOS structure's energy band diagram at $V_G = V_{FB}$ the energy bands are horizontal (flat) not only in the bulk, but also at the surface.
MOS

flexible display an emissive display based on organic LEDs or quantum nanodot LEDs that can be flexed/bent without altering its operation.
organic LED

flexible electronics and photonics semiconductor devices and circuits formed on the flexible substrates, e.g. flexible displays, flexible LED-based lighting panel, solar cells, thin-film transistors arrays, etc., compatible with wearable electronics and photonics and bionic skin.
wearable electronics and photonics

flexible substrates the substrates on which flexible semiconductor electronic and photonic devices are formed; in most cases plastic sheets or rolls, but also various types of fabrics.
flexible display, wearable electronics/photonics

flicker noise a noise common in semiconductor devices; a low-frequency distortion of the signal; manifest itself as a resistance fluctuation which is then reflected in the voltage or current fluctuations.
noise

flip chip technology IC assembly method in which chip is attached directly to the printed board with its surface down; chip is electrically connected to the package via an array of solder balls and without wire bonding; chip has appropriately positioned and pre-processed bond pads.
Ball Grid Array, wire bonding, assembly

flip chip bonding see *flip chip technology*.

float-zone crystal growth, FZ a method used to form single-crystal semiconductor substrates; alternative to CZ (Czochralski) crystal growth process; polycrystalline material (typically in the form of a circular rod) is converted into single-crystal by zone heating (zone melting) starting at the plane where single crystal seed is contacting polycrystalline material; used to make Si wafers; results in very high purity (i.e. very high

resistivity) single crystal Si; does not allow as large Si wafers as CZ does; diameter of FZ wafers up to 150 mm; radial distribution of dopant in FZ wafer is not as uniform as in CZ wafer; wafers used in high-end Si microelectronics are almost uniquely CZ grown; FZ wafers are used to make high power devices.
Czochralski crystal growth, Bridgman method

FLP see *Fermi Level Pinning*.

fluorescence emission of the longer wavelength (lower energy) light from certain materials stimulated by their exposure to the shorter-wavelength (higher energy) radiation (e.g. UV or X-ray); results from the electrons transitioning to the higher energy state by absorbing energy of short-wavelength radiation and then releasing energy by the emission of longer-wavelength light; a form of luminescence.
luminescence, photoluminescence, electroluminescence

fluorinated oxide addition of fluorine reduces dielectric constant k of SiO_2 from 3.9 to 3.5 making it better suitable for low-k dielectric application as an interlayer dielectric in high-end ICs; on the other hand fluorine in SiO_2 facilitates undesired boron penetration by modifying oxide's density.
interlayer dielectric, low-k dielectric

fluorine, F_2 a halogen gas; highly reactive and very toxic; fluorine is an important component of the etching chemistries both wet and dry in Si processing; e.g. gases: NF_3, CF_4, and several others; liquids: HF-water solution (etchant of SiO_2); fluorine in etching chemistries catalyzes breaking of Si-Si an Si-O bonds and promoting etching.

fluorine, F_2, excimer laser wavelength of emitted radiation 157 nm; would be highly suitable for the exposure of sub-10 nm patterns if not for the low maximum output power of 6 mJ/pulse.
excimer laser

fluorine passivation saturation of broken bonds on the Si surface with fluorine (e.g. by immersion in HF-water solution) renders Si surface chemically inactive; used also to passivate surfaces of stainless steel, nickel etc.
surface passivation

F-N tunneling see *Fowler-Nordheim tunneling.*

forbidden band see *energy gap (bandgap).*

forming gas a mixture of hydrogen (not exceeding 10%) in nitrogen; used as an ambient in annealing processes used in semiconductor manufacturing.

forward bias a bias at which potential barrier at the *p-n* or metal-semiconductor junction is lowered and the large current is allowed to flow from one region to another.
reverse bias, potential barrier

FOUP Front Opening Unified Pod; a SMIF pod used in advanced semiconductor manufacturing.
SMIF, SMIF pod

four-point probe the most common techniques for measuring resistivity and the doping level of semiconductors; uses four equally spaced metal probes touching semiconductor; two outside probes are used to flow the current in semiconductor while two inside probes measure resulting voltage drop which is proportional to semiconductor resistivity; a doping level of semiconductor wafer is determined from the four-probe resistivity measurements; a technique broadly used to monitor doping processes in semiconductor device manufacturing.
resistivity, doping level

Fourier-Transform Infrared Spectroscopy, FTIR a method used to investigate composition of solids on the basis of analysis of spectral absorption bands in the infrared regime; uses Fourier transform spectrometer; samples must be transparent to infrared radiation.
infrared radiation

Fourier-Transform Photoluminescence Spectroscopy, FTPL very useful in characterization of epitaxial layers; shows high sensitivity in detecting dopants in semiconductors.
epitaxial layer

Fowler-Nordheim tunneling, F-N the current flowing across MOS structure at the high electric field in the oxide; electrons tunnel from semiconductor conduction band into the oxide conduction band through

part of the potential barrier at the semiconductor-oxide interface; most likely to dominate leakage current in MOS structures with oxide 5-10 nm thick.
tunneling, direct tunneling, leakage current

FRAM   same as FeRAM; Ferroelectric Random Access Memory.
RAM

free (charge) carriers   electrons in the conduction band and holes in the valence band of semiconductor which are free to move and to carry electric charge and to contribute to semiconductor electrical conductivity.
free electron, free hole

free electron   any electron in the conduction band that is not bound to an ion, atom, or molecule, and hence, is free to move around carrying electric charge.
electron

free hole   any hole in the valance band that is not bound to an ion, atom, or molecule, and hence, is free to move around carrying positive electric charge.
hole

freeze-out   a state of the semiconductor material occurring at the temperatures too low to initiate dopant ionization (activation) process; dopants remain electrically neutral, i.e. do not generate free electrons and holes.
dopant activation, ionization energy

Frenkel defect   a crystal defect; a combination of interstitial and vacancy; formed when an atom is displaced from its lattice site to an interstitial site; can migrate through the crystal.
crystal defects, point defects, interstitial defect, vacancy

Frenkel pair   see *Frenkel defect.*

Front-End-of-Line, FEOL, processes   operations performed on semi-conductor wafer in the course of chip manufacturing up to the first metallization.
Back-End-of-Line processes

FSB see *Fixed Shape Beam*.

FSG Fluorine Doped Silicate Glass see *fluorinated oxide*.

FTIR see *Fourier-Transform Infrared Spectroscopy*.

FTPL see *Fourier Transform Photoluminescence Spectroscopy*.

full-field exposure an exposure of the entire wafer during photo-lithographic processes through the properly designed mask; used in contact and projection printing; as opposed to the step-and-repeat exposure process used in steppers.
contact printing, proximity printing, step-and-repeat projection

Full Wafer Imaging, FWI same as *full-filed exposure*.
step-and-repeat projection

fullerenes carbon molecules in the form of tubes, ellipsoids, or hollow spheres.
carbon nanotube

Fully-Depleted FinFET a FinFET with a "fin" sufficiently narrow to allow a full depletion of the transistor channel; essentially an implemen-tation of the *Fully-Depleted MOSFET* concept in the FinFET (vertical) configuration.
FinFET, depletion region, channel

Fully-Depleted MOSFET, FDMOSFET a MOSFET implemented in Ultra-Thin SOI or Extremely-Thin SOI substrates; at the extreme thinness of the silicon active layer (down to ~ 5 nm) its full depletion under the typical bias condition is possible; under those condition a transistor channel is a fully depleted/inverted silicon active layer; results in the improvements of the transistor characteristics; see also *DST, Depleted Substrate Transistor*.
ET SOI, active Si layer, MOSFET

Fully-Depleted SOI, FD SOI a class of SOI substrates featuring a silicon active layer so thin that it can be fully depleted; used in Fully-Depleted MOSFET technology (*see above*).
SOI, active Si layer

81

functional oxides a class of oxygen containing ternary compounds which display variety of complex emergent behaviors beyond the confines of the traditional properties of the oxides: same as complex oxides.
complex oxides

furnace a tool used in semiconductor device manufacturing to process wafers at high temperature in the ambient of strictly controlled composition; uses heavy heating coils, and hence, does not allow rapid changes of wafer temperature, i.e. is incompatible with low-thermal budget RTP (Rapid Thermal Processing); designed for high thermal budget batch processes.
batch process, Rapid Thermal Processing, thermal budget

furnace, horizontal see *horizontal furnace.*

furnace, vertical see *vertical furnace.*

FWI see *Full Wafer Imaging.*

FZ see *float-zone crystal growth.*

G

g-**factor** defined in the course of Electron Paramagnetic Resonance (EPR) measurements used in semiconductor characterization; provides information regarding the electronic structure of the paramagnetic centers (defects) detected by EPR.
EPR

g-**line photolithography** photolithography using 436 nm wavelength UV light for exposure; a high-intensity line at 436 nm in the spectrum of UV lamp is referred to as "*g*-line".
i-line lithography, high pressure Hg lamp

GAA Gate All-Around see *gate all-around.*

gadolinia see *gadolinium oxide Gd_2O_3.*

gadolinium oxide, Gd$_2$O$_3$ a rare earth metal oxide with a dielectric constant k in the range 10-16; features high dielectric strength and low leakage current; a candidate for high-k MOS gate dielectric applications; represents so-called 2nd generation high-k dielectrics.
high-k dielectric, gate dielectric

gallium   an element from the group III of the periodic table; forms some key III-V compound semiconductors with selected elements from the group V of the periodic table.
gallium antimonide, gallium arsenide, gallium nitride

gallium antimonide, GaSb   a III-V compound semiconductor; energy gap $E_g = 0.72$ eV, direct; zinc-blend crystal structure; mobility of electrons and holes at 300 K: 5000 and 850 cm^2/V-s respectively; of interest because at the high electron mobility; it also features relatively high hole mobility as compared to other III-V semiconductors.
compound semiconductor, AIII-BV, antimonides

gallium arsenide, GaAs   a III-V compound semiconductor, energy gap $E_g = 1.43$ eV, direct; zinc-blend crystal structure; mobility of electrons and holes at 300 K: 8500 and 400 cm^2/V-s respectively; thermally unstable above 600 °C due to As evaporation; does not form good quality native oxide, and hence, MOS devices are not possible with GaAs; due to the direct bandgap commonly used to fabricate light emitting devices in the 850-940 nm range; also foundation of the variety of high-speed electronic devices; bandgap can be engineered by forming ternary compounds based on GaAs, e.g. AlGaAs extends the emission spectrum to 650 nm.
compound semiconductor, AIII-BV, direct bandgap, arsenides

gallium arsenide-on-silicon, GaAs-on-Si   heteroepitaxial growth of GaAs on Si using a buffer layer (e.g. strontium titanate, SrTiO$_3$ - a dielectric which bonds to both the GaAs and Si and features lattice structure halfway between the two); GaAs-on-Si allows integration of Si based IC technology (electronic functions) with GaAs based photonic devices (photonic functions); makes available GaAs substrates as large as Si wafers at the cost much lower than the cost of bulk GaAs wafers.
heteroepitaxy, buffer layer, strontium titanate

gallium nitride, GaN a III-V compound semiconductor, energy gap $E_g = 3.4$ eV, direct; wurtzite crystal structure; mobility of electrons and holes at 300 K: 300 and 350 cm^2/V-s respectively; a wide-bandgap semiconductor capable of emitting short wavelength light in the violet-blue-green range; used to fabricated blue-green LEDs and lasers; foundation of the LEDs based lighting; also used for UV detection as well as in power electronics; due to the limited availability of the free standing single-crystal GaN wafers typically epitaxially deposited on sapphire or SiC substrates; bandgap can be engineered by forming ternary compounds based on GaN, e.g. InGaN covers the emission spectrum from about 360 nm to 550 nm.
laser, LED, power device, semiconductor lighting, white LED, nitrides

gallium phosphide, GaP a III-V compound semiconductor; energy gap $E_g = 2.26$ eV, indirect; zinc blend crystal structure; mobility of electrons and holes at 300 K: 110 and 75 cm^2/V-s respectively; with certain amount of As added the bandgap changes from indirect to direct; important component of the III-V ternary (GaAsP, AlGaP) light emitters (LEDs) covering the emission spectrum from 650 nm (red) to 550 nm (green).
LED, indirect bandpap; phosphides

Gas Chromatography Mass Spectroscopy, GS-MS in solid-state technology used to study species adsorbed on the solid surfaces (desorption + analysis by MS); in semiconductor practice a technique used to detect organic contaminants on the silicon surface.
organic contaminants, TD-GC-MS

gas-phase mass transfer a characteristic which controls the mass transfer dynamics in the gas-phase processes such as CVD.
CVD

gate in general, a terminal in semiconductor devices used to trigger current flow; one out of three terminals in the Field Effect Transistors; controls output current (i.e. flow of carriers in the channel); in MOSFET the gate is comprised of the gate contact and a thin dielectric; in MESFETs the gate is a Schottky contact; in JFET the gate is a *p-n* junction.
Field Effect Transistor, MOSFET, MESFET, JFET

gate all-around a MOS gate stack in which gate dielectric and gate contact are "wrapped around" the channel; possible with vertical MOSFET architectures such as FinFET.
gate stack, wrapped around gate, FinFET, omega gate, pi gate

gate capacitance for MOSFET /CMOS to operated properly the MOS gate stack must feature certain minimum capacitance $C \sim kA/d$; as gate contact area A decreases due to scaling, thickness of gate dielectric d must decrease to maintain desired capacitance of the stack; when reduction of gate dielectric thickness is not possible due to the excessive tunneling current the dielectric featuring dielectric constant k higher than that SiO_2 must be used in MOS gates.
alternative dielectric, high-k dielectric, tunneling

gate contact a conducting material (metal, poly-Si, or silicide) in the gate structure of various FETs.
gate stack, poly-Si gate, silicide, Field Effect Transistor

gate dielectric a very thin layer of an insulator sandwiched between semiconductor and the gate contact in MOS devices; in silicon MOSFETs it is typically a thermally grown SiO_2, often nitrided; depending on application it can be as thin as 1.0-1.2 nm (advanced digital integrated circuits) or as thick as 50 nm (discrete power MOSFETs); in ultra-small geometry (below 45 nm) CMOS ICs the SiO_2 is replaced with gate dielectrics featuring higher than SiO_2 dielectric constant k such as e.g. hafnium dioxide HfO_2; dielectrics such as Si_3N_4 and Al_2O_3 are used as gate dielectrics in non-silicon MOSFETs as well silicon and non-silicon TFTs.
high-k dielectric, silicon dioxide, MOSFET, nitrided oxide, thermal oxidation, Thin-Film Transistor

gate injection during the constant-current stress of the MOS gate stacks (oxide reliability test) electrons are injected into the oxide from the gate contact (negative bias no the gate contact).
constant-current stress, substrate injection

gate leakage a leakage current flowing across the MOS gate structure in the MOSFET; contributor to the possible overheating of the logic ICs.
gate dielectric, leakage current, MOSFET

gate length effective length of the gate contact at the surface of the Si substrate in the MOSFET in the direction of current flow; physical gate length is typically smaller than channel length; an approximate representation of the various technology nodes (technology generations).
gate scaling, technology generation

gate oxidation process of thermal oxidation of silicon which forms an oxide for the MOS gate stack; typically carried out in dry oxygen at the temperature in the range 700 °C - 1000 °C depending on the desired thickness of the gate oxide.
gate oxide, thermal oxidation, gate stack

gate oxide see *gate dielectric*.

gate oxide integrity, GOI term refers to electrical "integrity" of the gate oxide; determined through various current/voltage/electric field stress tests of the MOS gate stacks.
charge-to-breakdown, ramp voltage breakdown, SILC, TDDB

gate scaling process of the continued reduction of the gate length for the purpose of improving performance of the CMOS transistors comprising logic and memory ICs; a driving force behind continued improvement of the performance of silicon digital ICs since early 1970s; because of the physical barriers, manufacturing challenges, and cost alternative to gate scaling ways to continue progress in logic IC technology are pursued.
gate length, technology generation

gate self-aligned process key part of the planar MOSFET/CMOS fabrication sequence; gate stack is used as a mask during source and drain implantation, and hence, gate stack self-aligns with respect to the source and drain regions.
planar transistor

gate stack gate metal (conductor) - gate oxide (dielectric) structure in a MOSFET/CMOS.
MOS gate, gate oxide

Gaussian distribution often informally referred to as a "bell curve"; in semiconductor engineering describes distribution of implanted and driven-in dopants.
ion implantation, drive-in

GC-MS see *Gas Chromatography Mass Spectroscopy.*

GCA see *Gradual Channel Approximation.*

generation process of free charge carriers formation in semiconductor; in band-to-band generation electron acquiring energy in excess of the energy gap moves from the energy state in the valance band to the energy state in the conduction band leaving free hole in the valance band; formation of electron-hole pair results.
band-to-band generation, recombination

generation current current resulting from the generation of electron-hole pairs in the space charge region of the reverse biased *p-n* junction.
p-n junction, recombination current

generation lifetime an average time needed to generate an electron-hole pair in the space charge region of the junction.
recombination lifetime

generation rate the rate at which electron-hole pairs are generated (number/cm^3 sec).
recombination rate

generation-recombination current the current flow caused by the generation-recombination process of charge carriers in semiconductor.
generation, recombination

GeOI, or GOI see *Germanium-On-Insulator*

germanides alloys of germanium and metals; ohmic contact materials in germanium devices in the same way as silicides are used as contact materials in silicon device; e.g. nickel germanide NiGe.
ohmic contact, silicide

germanium, Ge an elemental semiconductor from the group IV of the periodic table; energy gap $E_g = 0.66$ eV, indirect; cubic crystal structure; mobility of electrons and holes at 300 K: 3900 and 1900 cm^2/V-s respectively both significantly higher than in Si; in contrast to Si does not form high quality native oxide, but works well with high-k dielectrics to form MOS/CMOS structures; also important in conjunction with

silicon as silicon germanium SiGe; a replacement for silicon in some applications.
high-k dielectrics, silicon germanium

Germanium-On-Insulator, GeOI or GOI same concept as SOI; due to the higher electron mobility in Ge than in Si, faster circuits are possible with GeOI substrates; also allows high-speed photodetectors which is of importance for the on-chip optical interconnects.
germanium, SOI, optical interconnects

germanium tin, GeSn group IV-IV semiconductor compound featuring direct energy gap $E_g = 0.8$ eV; demonstrated in semiconductor laser applications.
compound semiconductor, semiconductor laser

gettering process which moves contaminants and/or defects in semiconductor toward its bulk and away from its top surface and traps them there; creates "denuded zone" at the top surface.
denuded zone, extrinsic gettering, intrinsic gettering

GILD Gas Immersion Laser Diffusion.

global planarization a planarization process based on CMP resulting in flat surface across the entire wafer; as opposed to smoothing, partial planarization or local planarization; key element of the damascene process.
Chemical-Mechanical Planarization, smoothing, partial, local planarization, damascene process

global strain see *biaxial strain.*

glow discharge electrical discharge in gases.

GNR see *graphene nanoribbon.*

GOI see *Gate Oxide Integrity.*

GOI see *Germanium on Insulator.*

gown   protective garment used by personnel working in the cleanroom facilities; its main function is to prevent particles generated by the cleanroom personnel from penetrating cleanroom air.
cleanroom

gowning   process of replacing street clothing with cleanroom compatible gowns; gowning area is a dedicated part of the cleanroom facility.
gown

graded heterojunction   composition of the layer formed at the interface between two single-crystal semiconductors featuring different bandgaps is gradually modified to accommodate the transition.
heterojunction, buffer layer

graded junction   a dopant concentration in the p-n junction region is changed gradually (e.g. linearly) across the junction; used to control electric field within the junction region.
abrupt junction, linearly graded junction

Gradual Channel Approximation, GCA   assumes that in the MOSFET's channel voltage changes gradually from the drain to the source; used in the modeling of MOSFETs characteristics.
drain, source, MOSFET

grain   a small in volume piece of single-crystal material, i.e. a material in which crystallographic long-range order is preserved; polycrystalline solids consists of the randomly connected grains which vary in size depending on material and formation/deposition method.
grain boundary, polycrystalline material

grain boundary   a boundary between grains in polycrystalline material; grain boundary is a discontinuity of the material structure having an effect on its fundamental properties including conductivity and lifetime of charge carriers.
polycrystalline material

graphene   a single atom thick sheet of carbon bonded in the hexagonal honeycomb lattice; layers of graphene stacked on top of each other and kept together by the weak van der Walls forces form graphite; features excellent thermal, mechanical, and electrical properties including extremely high conductivity due to the electron mobility that can reach

200000 cm^2/V·s; with zero energy gap (E_g = 0) graphene is a semi-metal rather than semiconductor; lack of the energy gap diminishes its potential in digital electronics (e.g. in transistors for logic applications); may be potentially useful in ultra-high frequency analog devices and circuits; rolled into a cylinder, graphene forms a single-walled carbon nanotube; graphene can be epitaxially grown on silicon carbide or by the CVD process on a liquid metal matrix.
carbon nanotube, van der Waals force, molybdenum disulfide, phosphorrene, slilicene

graphene nanoribbon, GNR graphene in the form of the strips featuring width typically below 50 nm; electronic properties of GNRs depend on the configuration of carbon atoms on the edge of nanoribbon; armchair and zigzag GNRs are distinguished.
graphene, armchair GNR, zigzag GNR

Gummel-Poon model often used model (circuit simulator) of bipolar transistor.
Ebers-Moll model, hybrid-pi model

Gunn diode a semiconductor (typically III-V GaAs or InP) microwave diode based on the Gunn effect; the first solid-state source of microwave power; also known as Transferred-Electron Diode (TED).

Gunn effect an effect occurring in compound semiconductors; electron drift velocity is decreasing as electric field is increasing above certain critical value resulting in the negative resistance; a diode displaying Gunn effect can produce microwave oscillations of current.
Gunn diode

H

h, **Planck constant** 6.63 x 10^{-34} J-s.

h_c critical thickness; see *pseudomorphic material.*

hafnium oxide, HfO$_2$ a high-k dielectric replacing SiO$_2$ as a gate dielectric in silicon CMOS integrated circuits in 45 nm and below technology generations; dielectric constant $k \sim 22$; thermally stable up

to 700 °C; commonly deposited by Atomic Layer Deposition; in many aspects similar to zirconium oxide.
gate oxide capacitance, high-k dielectric, zirconium oxide, Atomic Layer Deposition

hafnium silicate, HfSiO$_4$ high-k dielectric that can be used as a gate dielectric in silicon CMOS integrated circuits; dielectric constant $k \sim 15$; combines characteristics of silicon dioxide, SiO$_2$, and hafnium dioxide, HfO$_2$; thermodynamically stable with silicon, but inferior to HfO$_2$ in terms of charge trapping; in many aspects similar to zirconium silicate.
gate oxide capacitance, high-k dielectric, zirconium silicate

half-pitch half the distance between the mid-widths of the two adjacent interconnect lines in an IC.

Hall effect magnetic field applied to semiconductor through which current is flowing causes path of the flowing electrons to curve and leads to the asymmetric distribution of charge density across the sample, and hence, generation of voltage across the sample; used to derive selected electrical characteristics of semiconductor, e.g. carrier mobility.
Hall mobility

Hall mobility charge carrier mobility determined using Hall-effect measurement.
Hall effect, mobility

Hall voltage the voltage generated in semiconductor as a result of the Hall effect.
Hall effect

halo implantation a low energy, low current ion implantation carried out at the large incident angle so that implanted dopants penetrate underneath the edge of the MOS gate stack; used in CMOS fabrication to suppress short-channel effects.
ion implantation, short-channel effects

handle wafer a wafer used as a temporary support in various wafer bonding applications.
SOI, wafer bonding, wafer debonding

handler platform   centrally located chamber in the cluster tool to which all process module are attached; a housing of the robot handling wafers in the cluster tool.
cluster tool

hard bake   thermal curing step performed in the lithographic process after resist developing and before etching; aimed at the strengthening of the resist structure prior to etching; temperature of hard bake varies depending on the resist, but is lower than 200 °C.
soft bake

hard breakdown   catastrophic, irreversible breakdown of the gate oxide in MOS devices; oxide remains permanently damaged after the electric field is removed.
breakdown, oxide breakdown, soft breakdown

HB LED   High Brightness Light Emitting Diode; a class of short wavelength LEDs developed specifically for lighting applications.
light emitting diode

HBT   see *Heterojunction Bipolar Transistor.*

HCl   see *hydrochloric acid.*

HCI   Heavily Charged Ions, see *multicharged ion.*
single-charge ion

HDIS   High Dose Implanted (resist) Strip; a process of removing photo-resist which served as a mask during high-dose ion implantation.
ion implantation, implantation dose, photoresist

HDP   see *high density plasma.*

HDP-CVD   High Density Plasma Chemical Vapor Deposition.
PECVD, high density plasma

heat management   the term refers to the heat generated in the course of semiconductor device/chip operation; it needs to be managed in order to prevent overheating and ultimately catastrophic damage of the device/chip; a key challenge in the multi-billion transistors logic chips.
logic circuit, critical temperature

heavily charged ion, HCI   see *multicharged ion.*

HEED   see *High-Energy Electron Diffraction.*

Heisenberg uncertainty principle   applied to semiconductors the principle states that simultaneous exact determination of the position and momentum of the electron is not possible.

helicon plasma   high-density plasma generated using helicon antennas.
high density plasma, inductively coupled plasma

HEMT   see *High Electron Mobility Transistor.*
HFET

Henry's law   amount of gas in the liquid is directly proportional to the pressure of the gas above the liquid surface; example of application in semiconductor processing: control of concentration of ozone dissolved in deionized water (ozonated water).
ozonated water, ozone

HEPA filter   High Efficiency Particulate Air filter; HEPA filters are used in cleanrooms; capable of removing 99.99% of all particles 0.3 μm and larger from circulated air.
cleanroom, ULPA filter

hetero-integration   an integration by stacking into a self-contained system chips which are functionally different and are manufactured using different processing sequences; e.g. face-to-face bonded processor and memory stack integrated into a single package; 3D alternative to the system-on-chip (SOC) approach.
SOC, three-dimensional integration

heteroepitaxy   epitaxial deposition process in which chemical composition of epi material is different than the chemical composition of the substrate; e.g. AlGaAs epi on GaAs substrate or SiGe epi on Si substrate; special consideration is given to the lattice mismatch between epi layer and the substrate.
buffer layer, graded junction, homoepitaxy, lattice mismatch

heterojunction a junction, typically *p-n*, formed by two different semiconductor materials (e.g. Si and SiGe, or GaAs and AlGaAs).
homojunction

Heterojunction Bipolar Transistor, HBT high-performance transistor structure; unlike conventional bipolar transistors built using more than one semiconductor material, hence "heterojunction"; takes advantage of the different bandgap of semiconductors used to form emitter, base and collector; III-V compounds based HBTs: e.g. *n*-AlGaAs/*p*-GaAs/*n*-GaAs; Si-based HBTs use $Si_{1-x}Ge_x$ material system in which *x* is varied to accomplish desired bandgap; HBTs are manufactured using high precision epitaxy such as MBE or MOCVD.
heteroepitaxy, bipolar transistor, Molecular Beam Epitaxy

Heterojunction Field Effect Transistor, HFET family of FETs built using epitaxially formed structure consisting of semiconductors featuring different bandgaps; typically based on modified GaAs (e.g. AlGaAs) material system.
gallium arsenide, heterojunction

hexagonal crystal system one out of seven unit cells (crystal systems) in crystallography; defined by four axes among which three are of the same length and lie in the same plane to which the fourth axis is perpendicular; among semiconductors some II-VI compounds (e.g. ZnO and CdSe) and III-V compounds (e.g. AlN, GaN) feature hexagonal lattice.
cubic system, wurtzite lattice

HEXFET Hexagonal cell FET.

hexode etcher dry etching tool which uses six-sided wafer holder for the purpose of increasing process throughput.
throughput

HF see *hydrofluoric acid*.

HF:HCl:H₂O a weak solution used in some silicon cleaning sequences as replacement for HPM and HF-last processes.
HPM, hydrochloric acid, hydrofluoric acid, HF-last

94

HF-last, HFL "HF-last" process; Si surface preparation sequence in which HF:H_2O etching of native/chemical oxide is performed at the end of the sequence leaving silicon surface hydrogen terminated.
hydrogen termination

HF vapor a hydrofluoric acid above 19 °C; mixed with water vapor or a vapor of an alcoholic solvent is a very effective isotropic etch of silicon dioxide; used extensively in MEMS release processes.
hydrofluoric acid, anhydrous HF, MEMS release

HFET see *Heterojunction Field Effect Transistor.*

HIGFET Heterojunction Insulated Gate Field Effect Transistor; type of the IGFET using heterojunction as a channel.
IGFET

high-angle implantation ion implantation process carried out at the high angle between the incident beam and surface normal.
ion implantation

high-current implanter an implanter designed to perform implantation at high ion beam current.
ion implantation, implanter, ion beam current

high-density plasma, HDP plasma featuring high concentration of free electrons, and hence, high concentration of ions; denser the plasma, higher the efficiency of the process (etching, deposition); in general, the HDP is accomplished through the magnetic confinement of plasma.
plasma, magnetically confined plasma, ECR plasma, helicon plasma, ICP

High Electron Mobility Transistor, HEMT a transistor formed using different III-V compounds to exploit difference in their bandgaps to increase mobility of electron; belongs to HFET device family.
HFET

High Energy Electron Diffraction, HEED a method used in characterization of physical features (crystallographic structure) of solids; by increasing energy of electrons impinging on solid surface additional information regarding material structure can be revealed.
LEED. RHEED

high-frequency C-V capacitance-voltage measurement carried out at the frequency of 100 kHz and higher (often 1 MHz); in MOS gate characterization often carried out in conjunction with low-frequency (quasistatic) C-V measurement.
capacitance-voltage measurements, quasistatic C-V

high-k dielectric a dielectric material featuring dielectric constant k higher than 3.9 which is k of SiO_2; used as gate dielectric in MOS devices and in storage capacitors; high k increases capacitance, or keeps it unchanged at the reduced area of the MOS gate allowing gate dielectric sufficiently thick to prevent excessive tunneling current.
alternative dielectric, gate capacitance, gate oxide, low-k dielectric, tunneling current

high-pressure mercury (Hg) lamp a mercury lamp featuring high pressure of mercury vapor; source of UV radiation with high-intensity wavelengths in the 350 nm - 500 nm regime; used for the photoresist exposure in g-line and i-line photolithographies.
photoresist, low pressure mercury lamp, g-line, i-line photolithography

High-Pressure Oxidation, HIPOX thermal oxidation of silicon carried out at the pressure of oxidizing ambient significantly higher than atmospheric pressure (e.g. 25 atm.); allows fast growth of thick oxide at reduced temperature; capable of enforcing oxidation of silicon nitride covered silicon resulting in nitrogen-rich layer near the surface of the grown oxide.
thermal oxidation, low pressure oxidation

hillock in semiconductor terminology the term refers to the local accumulation of metal resulting from the electromigration occurring in the interconnect lines.
electromigration, interconnect

HIPOX see *High Pressure Oxidation*.

HKMG High-k (dielectric) Metal Gate; also referred to as HK+MG; refers to the advanced (below 45 nm technology generation) MOSFET/ CMOS process which uses high-k dielectric instead of SiO_2 as a gate dielectric and metal gate contact instead of poly-Si.
gate contact, gate dielectric

HMDS Hexamethyldisilizane; an adhesion promoter; material improving adhesion of photoresist to the surface; deposited on the wafer surface prior to photoresist deposition.
adhesion

hole a positive charge carrier in semiconductors which materially does not exist; hole is a lack of electron moving in the direction opposite to that of electron and carrying positive charge; features higher effective mass than electron, hence, lower mobility.
electron, electron-hole pair, hole mobility

hole mobility parameter which is a measure of scattering of the hole moving in semiconductor; unit cm^2/Vsec; proportionality factor between hole drift velocity and electric field as well as conductivity and hole concentration in semiconductor; due to the higher effective mass of a hole, hole mobility is typically significantly lower than electron mobility; hole mobility at 300 K for: Si - 500 cm^2/Vsec; GaAs - 320 cm^2/Vsec, GaN - 350 cm^2/Vsec, Ge - 1800 cm^2/Vsec, 6HSiC - 100 cm^2/Vsec.
electron mobility, mobility, scattering, drift velocity

hole scattering see *scattering*.

homoepitaxy an epitaxial deposition process in which chemical composition of epi material and the substrate are the same; e.g. Si epi layer on Si substrate.
epitaxy, heteroepitaxy, epi layer

homojunction a junction formed of one material, however, the two parts of the junction display different properties (e.g. conductivity type in the case of *p-n* junction or doping concentration in the case of low-high junction).
heterojunction, low-high junction, p-n junction

homopolar bond a bond between the atoms of the same polarity; covalent bond is homopolar.
covalent bond

hopping current conduction mechanism in which charge carriers are "hopping" from defect to defect.
defect

horizontal furnace a furnace for high temperature processing of semiconductor wafers in which process tube is positioned horizontally and wafers are located in the boat vertically on their edges; common furnace configuration in semiconductor manufacturing; used for thermal oxidation, CVD, diffusion and anneals; batch processor; an alternative configuration to vertical furnace; inferior to vertical furnace in terms of the foot print, wafer loading automation and heating uniformity.
batch process, vertical furnace

hot carrier diode term used in reference to a Schottky diode.
Schottky diode

hot electron an electron which is not in thermal equilibrium with the lattice; typically carrier "heating" occurs in the regions of semiconductor device featuring very high electric field; hot electrons create reliability problems in semiconductor devices.
short-channel effects

hot plate a heated plate used in semiconductor processing to cure photoresist and to carry out other low-temperature (< 300 °C) anneals in ambient air.
soft bake

hotwall reactor thermal reactor using radiant heating; heating elements are located outside of the process chamber, and hence, walls of the process chamber are in thermal equilibrium with wafers located inside.
coldwall reactor

HPM Hydrochloric acid-hydrogen Peroxide-water Mixture; typically 1:1:5; applied at 40 °C to 70 °C; same as SC-2 and RCA-2; a cleaning solution used primarily to remove metallic contaminants; gradually replaced with alternative recipes such as those involving very weak solutions of HF:HCl in water, or abandoned.
metallic contaminant, wet cleaning, RCA clean

HREM High Resolution Electron Microscopy; electron microscopy techniques allowing atomic resolution imaging.
scanning electron microscopy

HRTEM High Resolution Transmission Electron Microscopy.
transmission electron microscopy

HSQ Hydrogen Silsesquioxanes; inorganic low-k dielectrics; dielectric constant $k \sim 2.9$, used as an interlayer dielectric in advanced multilayer interconnect schemes.
ILD, low-k dielectrics

HVAC High Vacuum.

hybrid clean a cleaning process which combines wet (liquid-phase) and dry (gas-phase) steps, e.g. UV/Ozone followed by HF:H_2O.
dry cleaning, wet cleaning, hydrofluoric acid, UV/ozone

hybrid IC an electronic circuit integrated on the ceramic substrate using various components fabricated using diversified technologies, e.g. monolithic, thick film, etc. and then enclosed in the single package; substrate does not participate in the operation of the circuit; common approach in the implementation of microwave ICs.
monolithic IC, thick film, radio frequency, microwave IC

hybrid-*pi* model a basic model (equivalent circuit) describing operation of bipolar transistor primarily in amplification applications.
Ebers-Moll model, Gummel-Poon model

hydrochloric acid, HCl used in cleaning solutions (HPM, or HF:HCl:H_2O, mixture) to complex metallic contaminants on the silicon surface; gaseous HCl may be added to oxygen during thermal oxidation of Si for the same purpose.
cleaning, HPM

hydrofluoric acid, HF extremely hazardous and corrosive acid commonly used in silicon processing to etch silicon dioxide, SiO_2; also, important component of the essentially all Si cleaning recipes and some III-V semiconductors cleaning recipes; a colorless liquid at atmospheric pressure below 19 °C; a gas above 19 °C; commercially available as anhydrous HF and aqueous HF (49% solution).
anhydrous HF, buffered oxide etch, DHF

hydrogen peroxide, H_2O_2 a liquid; very strong oxidizer; common component of semiconductor cleaning solutions; in some cases replaced by ozone.
APM, HPM, ozone

hydrogen plasma discharge in hydrogen; non-thermal equivalent of hydrogen reduction of surface oxides; commonly used in surface conditioning of III-V compounds in which high-temperature hydrogen reduction processes cannot be used.
plasma, hydrogen reduction

hydrogen reduction a reduction (removal) of native oxide from the Si surface during annealing in H_2 ambient at the temperatures exceeding 900 °C; procedure often used to remove *in situ* native/chemical oxide from the Si surface prior to epitaxial deposition of silicon.
chemical oxide, native oxide

hydrogen termination broken silicon bonds at the surface are saturated (passivated) with hydrogen; Si-H is a stable bond preventing oxidation of silicon surface at room temperature; best accomplished by dilute $HF:H_2O$ etch followed by limited water rinsing.
HF last

hydrogenated amorphous silicon an amorphous silicon (a-Si) containing substantial quantities of hydrogen; hydrogen passivates Si dangling bonds and results in significantly improved electrical properties of a-Si; commonly used to fabricate thin-film silicon solar cells.
solar cell, amorphous Si

hydrophilic surface a surface tension is such that water is wetting the surface; contact (wetting) angle is close to 0°; e.g. oxide covered Si surface.
surface tension, wetting angle, hydrophobic surface

hydrophobic surface a surface tension is such that water is bidding on the surface; contact (wetting) angle is close to 90°; e.g. oxide-free Si surface.
surface tension, wetting angle, hydrophilic surface

I

i **- line photolithography** photolithography using 365 nm wavelength UV light for exposure; high-intensity line at 365 nm in the spectrum of UV lamp is referred to as "*i*-line".
g-line photolithography, high pressure Hg lamp

I-V measurements see *current-voltage measurements.*

IBD Ion Beam Deposition, a PVD thin-film deposition technique.
ion beam, PVD

ICP see *Inductively Coupled Plasma.*

ICP MS see *Inductively Coupled Plasma Mass Spectroscopy.*

ideal MOS an "ideal" MOS (Metal Oxide Semiconductor) is a structure for which it is assumed that: *(i)* there is no work function difference between metal gate and semiconductor, *(ii)* there is no charge at the semiconductor-oxide interface, and *(iii)* there is no charge in the oxide; all this means that with no voltage applied to the gate, energy bands in the MOS structure are horizontal, or flat; any departure from the "ideal" condition will change potential distribution across the structure.
flat-band voltage, oxide charge, work function

ideal *p-n* junction a *p-n* junction diode operation of which is described by the current-voltage relationship in which an ideality factor $n = 1$.
ideality factor

ideal Schottky contact a metal-semiconductor contact in which potential barrier is defined solely by the difference in metal and semiconductor work functions; i.e. there is no effect of interface states on the characteristics of a Schottky contact.
potential barrier, work function, work function difference

ideality factor a parameter (n) in the diode current-voltage relationship; close to 1 when carrier diffusion dominates current flow; approaches 2 when recombination current dominates.
p-n junction

IGBT see *Insulated Gate Bipolar Transistor.*

IGFET Insulated Gate Field Effect Transistor, effectively a MOSFET.
MOSFET

IGZO Indium (In) Gallium (Ga) Zinc (Zn) Oxide (O); transparent compound semiconductor; used to manufacture Thin-Film Transistor

(TFT) backplanes in LCD displays featuring higher resolution than displays using amorphous-Si TFT backplanes.
Thin-Film Transistor

I²L   see *Integrated Injection Logic.*

ILD   see *Inter-Layer Dielectric.*

imaging resist   a layer of photoresist in which photochemical reactions resulting from illumination are taking place; as the wavelength of exposure light is getting shorter, and hence absorption coefficient increases, thickness of the imaging resist is gradually reduced.
multilayer resist, planarizing resist

IMD   Inter-Metal Dielectric see *Inter-Layer Dielectric.*

IMEC   Interuniversity Microelectronics Center, Leuven, Belgium; founded in 1984; among international leaders in semiconductor research and development.

IMEC clean   wet cleaning sequence developed at IMEC; includes three cleaning steps: SOM+APM+dHF/HCl with DI water rinses in between and Marangoni drying at the end.
APM, Marangoni drying, SOM, deionized water

immersion cleaning   cleaning process in which wafers in the cassette are completely immersed in the cleaning solutions; as opposed to spray cleaning or spin cleaning.
spin cleaning, spray cleaning, wet bench

immersion photolithography   photolithography techniques in which space between final projection lens and the wafer in the exposure tool is filled with water rather than air; use of medium featuring higher refractive index n ($n = 1$ for air while $n = 1.34$ for water) increases numerical aperture (NA) of the photolithography tool, and hence, increases resolution of the pattern transfer process.
numerical aperture, refractive index

impact ionization   generation of electron-hole pairs in semiconductor due to the collisions of the high kinetic energy species with the atoms in the lattice.
electron-hole pair, generation

IMPATT diode   Impact Ionization Avalanche Transit-Time diode; GaAs based high-performance microwave transit-time diode.
transit - time diodes

implantation   in semiconductor terminology synonymous with *ion implantation.*

implantation angle   an angle at which implanted ions are imping on the surface; defined and monitored for the purpose of *(i)* preventing ion reflection from the surface, *(ii)* minimization of the effect of channeling.
ion reflection, channeling

implantation damage   during ion implantation accelerated ions collide with atoms in the target material and displace them from their original lattice sites causing damage; after implantation crystallographic order of the implanted material is being restored by a brief anneal (typically RTA) at the temperature of 800 °C or higher.
ion implantation, rapid thermal processing, post-implantation anneal

implantation dose   a number of implanted ions per unite area (atoms/cm^2); proportional to ion beam current and implantation time; typical doses are in the range from 10^{11} to 10^{16} atoms/cm^2; dose determines concentration of implanted ions, and hence, doping level.
ion implantation, ion beam current

implantation energy   kinetic energy of implanted ions established through ions acceleration in the acceleration column of an ion implanter; determines depth of penetration of the implanted material by the ions, and hence, depth of the junction formed; quantitatively expressed by projected by the range R_p.
projected range, channeling

implanted junction   a *p-n* junction formed by means of ion implantation.

implanter   a tool used to carry our ion implantation processes.
ion implantation

impurity   in semiconductor terminology this term has two meanings; may be used in reference to either dopant or a contaminant.
contaminant, dopant

in situ **doping** dopant atoms are introduced into semiconductor during its growth; most commonly during the epitaxial growth of the single-crystal semiconductor.
epitaxy

in situ **monitoring** monitoring of the semiconductor manufacturing processes inside the process reactor; often in real time while the process is underway; as opposed to process monitoring based on the outcome of the process.

index of refraction, *n* see *refractive index.*

indirect bandgap the energy gap (bandgap) in semiconductor is configured such that the bottom of the conduction band and the top of the valence band do not coincide at the momentum $k = 0$, i.e. are shifted along the k axis with respect to each other.
direct bandgap, photon, phonon

indirect bandgap semiconductor a semiconductor featuring indirect bandgap; in the case of an indirect bandgap semiconductors the energy realized during band-to-band electron-hole recombination is predominantly in the form of a phonon rather than electromagnetic radiation (photon), hence, indirect bandgap semiconductors, e.g. Si, Ge, SiC, are not used to fabricate light emitting diodes (LEDs).
direct bandgap semiconductor, phonon, light-emitting diode

indirect recombination an electron-hole recombination via centers (energy levels) present in the bandgap of semiconductor rather than via direct band-to-band recombination; centers in the bandgap result from the crystal defects and/or contamination of semiconductor material.
recombination, direct recombination

indium, In an element from the group III of the periodic table; component of the III-V semiconductors (InSb, InP, InAs); can be used as a *p*-type dopant (acceptor) in silicon in alloyed *p-n* junction formation; do not used as a dopant in conjunction with ion implantation and diffusion doping processes.
alloyed junction, indium antimonide, indium arsenide, indium phosphide

indium antimonide, InSb a III-V compound semiconductor; energy gap $E_g = 0.17$ eV, direct; zinc-blend crystal structure; mobility of

electrons and holes at 300 K: 80000 cm^2/V-s and 450 cm^2/V-s respectively; the narrowest bandgap among practical semiconductors, but featuring the highest electron mobility; very high electron mobility makes it potentially attractive in digital applications, but a very narrow energy gap is a limitation; used in infrared emitters and detectors.
compound semiconductor, infrared emitter, - detector, antimonides

indium arsenide, InAs a III-V compound semiconductor; energy gap $E_g = 0.36$ eV, direct; zinc-blend crystal structure; mobility of electrons and holes at 300 K: 22600 cm^2/V-s and 200 cm^2/V-s respectively, narrow bandgap, high electron mobility material; wafers up to 300 mm in diameter are commercially available; used in infrared emitters and detectors.
compound semiconductor, infrared emitter, - detector, arsenides

indium phosphide, InP a III-V compound semiconductor; energy gap $E_g = 1.35$ eV, direct; cubic crystal structure; mobility of electrons and holes at 300 K: 4500 cm^2/V-s and 100 cm^2/V-s respectively; wafers up to 100 mm in diameter are commercially available; used to produce semiconductor infrared emitters and detectors as well as in high-speed electronics.
compound semiconductor, infrared emitter, - detector, phosphides

indium tin oxide, ITO a ternary compound; an optically transparent and electrically conductive oxide; energy gap $E_g \sim 4$ eV; very broadly used in semiconductor photonic devices as a transparent contact.
photonic device

inductive heating heating technique in which energy of *AC* signal (typically in RF range) is coupled directly to the object to be heated; temperature of this object is increased due to the Joules effect; coldwall reactors; same principle as in the conventional microwave oven; in semiconductor manufacturing less common now than it was in the past because of the insufficient heating uniformity of very large wafers.
coldwall reactor, RF, resistance heating

Inductively Coupled Plasma, ICP a high density plasma used in semiconductor processing; etching in particular; inductive, as opposed to capacitive, coupling allows electrodes to be located outside of the

reactor; the most common high-density plasma mode in semiconductor manufacturing.
high-density plasma

Inductively Coupled Plasma Mass Spectroscopy, ICP MS a method allowing identification of species adsorbed on the solid surface; commonly used to detect organic contaminants on semiconductor surfaces which are thermally desorbed and then identified by analyzing plasma emission spectrum.
organic contaminants, GC-MS

inelastic collision a collision during which there is a loss of kinetic energy, i.e. part of the kinetic energy is converted into the other form of energy (most commonly heat) and the momentum is not conserved; collisions between species in solids are almost always inelastic.
elastic collision

infrared detector a detector of infrared radiation; a diode made using semiconductors featuring energy gap E_g < 1.7 eV; typically ternary compounds with tunable E_g formed on the basis of binary narrow-bandgap semiconductors, e.g. InGaAs; also II-VI compounds such as CdHgTe, and others; used in thermal imaging devices and household remotes.
infrared radiation, photodetector

infrared emitter an emitter of infrared radiation; a LED manufactured using semiconductors featuring direct bandgap E_g < 1.7.eV; between others used in household remotes.
infrared radiation, light emitting diode

infrared radiation, IR wavelengths from ~0.7 μm to ~15 μm range; invisible at the longer wavelengths in this range; used in night vision systems and household remotes; also used for heating purposes.
light emitter, photodetector

ingot circular in cross-section, elongated piece of single-crystal semi-conductor material typically resulting from the Czochralski crystal growth process; ingot is machined into the desired diameter and then sliced, using high-precision diamond saw, into wafers used to manu-facture semiconductor devices.
Czochralski crystal growth, single crystal, wafering

106

inject printing a method of patterned deposition using liquid precursors, e.g. organic semiconductors.
Physical Liquid Deposition

inorganic semiconductors include elements from the group IV of the periodic table: carbon, (C), silicon (Si), germanium (Ge) and tin (Sn); inorganic compound semiconductors can be formed by combining elements from the group IV (AIV-BIV), e.g. SiGe and SiC, elements from groups III and V (AIII-BV), e.g. GaAs, InP, or GaN, and elements from groups II and VI (AII-BVI), e.g. CdTe, ZnO, etc.; silicon is the most abundant and the most widely used inorganic semiconductor.
semiconductors, elemental semiconductors, organic semiconductors

Insulated Gate Bipolar Transistor, IGBT designed for power applications; device which combines MOS gate control and bipolar current flow mechanism; features high current/high voltage operation and high input impedance at the same time.
power device

insulator a very high resistivity material; in semiconductor device manufacturing used to electrically separate conducting materials; in general terms a type of solid which in contrast to metals and semiconductors does not conduct electrical current; insulators feature very wide energy gap E_g, typically above 5 eV; electrical conductivity of insulators cannot be altered by introduction of alien elements; the most common insulators in semiconductor technology: SiO_2, Si_3N_4, also high-dielectric constant k and low-k dielectrics.
metal, semiconductor, silicon dioxide, silicon nitride, dielectric constant

integrated bipolar transistor to meet specific requirements of the layout of an integrated circuit collector contact in an integrated bipolar transistor is located on the top surface of the wafer along with emitter and base contacts; in the discrete bipolar transistor collector contact is located on the back surface of the wafer; this change enforces changes in a configuration of the entire collector region of the bipolar transistor which includes buried n^+ region.
bipolar transistor, discrete transistor

integrated circuit, IC the entire fully functional electronic circuit formed on the single piece of a solid substrate and enclosed in a hermetic package; package is equipped with leads needed to electrically integrate

IC with a larger electronic system; depending on the substrate monolithic and hybrid ICs are distinguished.
hybrid IC, monolithic IC

integrated device a semiconductor device (e.g. transistor) designed to meet specific requirements of the circuit integration; as opposed to discrete device.
discrete device

Integrated Injection Logic, I^2L a logic gate based on bipolar transistor; commonly referred to as "*I*-square-*L*".
logic gate, ECL

integrated processing processing scheme employed in semiconductor manufacturing in which two or more consecutive operations are performed on the wafers within an enclosed, self-contained, multi-chamber reactor in the fully controlled ambient; tool implementing integrated process are commonly referred to as "clusters".
cluster

Inter-Layer Dielectric, ILD same as Inter-Level Dielectric and Inter Metal Dielectric; a dielectric material used to electrically separate closely spaced interconnect lines arranged in several levels (multilevel metallization) in an advanced integrated circuit; ILD must feature low dielectric constant k (as close to 1 as possible) to minimize capacitive coupling ("cross talk") between adjacent metal lines.
low-k dielectric, multilevel interconnect

Inter-Metal Dielectric see *Inter-Layer Dielectric*.

interconnect a conductor (typically metal) line connecting elements of an integrated circuit (typically transistors); in very high density integrated circuits interconnect lines form a multilevel network; in advanced silicon ICs interconnect lines are made out of copper.
copper, inter-layer dielectric, multilevel interconnects

interconnect density a measure of complexity of the multi-level metallization scheme in advanced ICs; expressed in $m/cm^2/level$.
multilevel interconnects

interdifussion an exchange by diffusion of atoms between two materials in physical contact in the presence of the concentration gradient and elevated temperature.
difussion

interface a region featuring finite thickness in between two materials in physical contact; a region within which chemical and/or structural transition between two materials occurs; typically highly disturbed region with potentially strong effect on the transport of charge carriers, and hence, device performance; e.g. carriers in the MOSFET channel flowing along semiconductor-dielectric (oxide) interface are subject to enhanced scattering.
bulk, surface, silicon-silicon dioxide system

interface state the allowed energy state within the energy gap at the interface between oxide (or metal) and semiconductor; essentially a surface state located within the interface region; an interface trap is an electrically active manifestation of the presence of an interface state.
surface state, interface trap

interface trap an electrically active interface state; a defect located at the interface between oxide and semiconductor; or metal-semiconductor; commonly associated with "dangling" (unsaturated) bonds at the semiconductor surface; capable of trapping and de-trapping charge carriers; interface traps have an adverse effect on device performance; in the case of free surface the same feature would be referred to as surface state.
dangling bond, interface state, trap

interfacial oxide an ultra-thin (< 1 nm) native oxide sandwiched between silicon and high-k dielectric in HKMG MOS gate stacks.
HKMG, MOS gate

internal emission the emission of electron from the valence band to the conduction band in semiconductor.
electron-hole pair

internal quantum efficiency, solar cells the efficiency with which photons penetrating solar cell, i.e. not reflected or transmitted out of the cell, generate charge carriers.
external quantum efficiency

interstitial an alien atom in the crystal lattice located in between lattice host atoms.
substitutional, vacancy, point defect

interstitial diffusion a diffusion mechanism in which diffusant moves in between host atoms in the lattice; low activation energy process; interstitial diffusants (e.g. gold in silicon) feature high diffusion coefficient.
diffusant, diffusion, substitutional diffusion

intrinsic concentration n_i concentration of charge carriers in the intrinsic, i.e. undoped, semiconductor; $n_i^2 = np$ where n and p represent concentration of free electrons and holes respectively.
intrinsic semiconductor

intrinsic delay the term refers to the logic gate; a delay internal to the gate which in other words means a gate delay.
logic gate

intrinsic Fermi level the Fermi level in an intrinsic semiconductor; located not exactly in the middle of the energy gap because of the different effective mass of an electron and a hole.
effective mass, Fermi level

intrinsic gettering a process in which gettering of contaminants and/or defects in semiconductor wafer is stimulated by the series of heat treatments applied in a strictly executed sequence and without any external interactions with a wafer; works well in forcing precipitation of oxygen in CZ grown single-crystal Si wafer away from its surface.
gettering, oxygen in silicon, oxygen precipitation, extrinsic gettering, CZ crystal growth

intrinsic semiconductor an undoped semiconductor material, i.e. semiconductor which does not display either n- or p-type conductivity; free electrons and holes result only from band-to-band generation, and hence, feature the same concentration ($n = p$); intrinsic concentration of carriers depend on the energy gap (bandgap); higher in narrow-bandgap and lower in wider bandgap semiconductors.
extrinsic semiconductor

inversion   an inversion of the conductivity type (from *p*- to *n*-type or vice versa) at the semiconductor surface in MOS structure induced by the electric field; in inversion the concentration of minority carriers exceeds that of the majority carriers; the effect of inversion is a foundation of the operation of the MOSFET.
accumulation, depletion, inversion layer, strong inversion, field effect

inversion layer   a layer at the semiconductor surface in the MOS structure in which condition of inversion is established; an inversion layer formed between the source and the drain of the MOSFET comprises a channel of the transistor; when surface is inverted a channel is created, and hence, the MOSFET is "on".
channel, MOSFET

inverter   see *CMOS inverter.*

I/O   input/output; a number of I/O pins in an IC package; reflects complexity of the circuit, i.e. higher the I/O count, more complex is the circuit.
DIP, PGA

ion   an ionized atom, i.e. atom deprived of at least one of its electrons to become positively charged, or atom acquiring additional electron(s) to become negatively charged.
ionization, ion multicharged, ion single-charge

ion beam   a geometrically confined and accelerated by electrostatic forces stream of ions directed toward desired target.
ion implantation, ion milling

ion beam current   expressed in terms of the number of ions crossing unit area in the unit time; a key parameter determining implantation dose in the process of ion implantation; a parameter used as a criterion distinguishing between various types of implanters.
ion implantation, implantation dose, implanter, high-current implanter

ion beam lithography, IBL   a lithography technique in which resist is exposed by accelerated ions; direct write or masked lithography; due to the limited scattering of ions in the resist the IBL may offer resolution of the pattern transfer process comparable to, or better than e-beam lithography; not used in commercial manufacturing due to the complexity of the ion optics infrastructure.
e-beam lithography, proximity effect, direct write lithography

ion implantation a process in which ions accelerated toward the solid surface penetrate the solid up to certain depth determined by the kinetic energy of impinging ions; in semiconductor technology used to introduce dopants, to form buried layers, and to modify solid surfaces; the most common selective doping technique in semiconductor device manufacturing; concentration of implanted dopant atoms depends on the ion dose while depth of implantation is determined by the kinetic energy of ions impinging on the surface.
doping, dose, implanter, projected range, selective doping

ion implantation damage see *implantation damage.*

ion milling subtractive process in which high energy ions impinging on the surface of the solid cause ejection ("sputtering") of the host atoms; highly anisotropic, non-selective and surface damaging etching.
etching, physical etching, sputtering, selective etch, anisotropic etch

ion, multicharged, MCI an atom ionized to a q ion charge level $q > 1$; e.g. Ar^{8+}; high power, very high frequency generators are needed to ionize gas to ion charge > 1; decelerated multicharged ions feature potential energy orders of magnitude higher than their kinetic energy; capable of interacting with solid surfaces at ultra-low velocities; in contrast to non-selective interactions of single-charge ions the multi-charged ions interact differently with insulators and conductors.
ion single-charge

Ion Projection Lithography, IPL modification of the ion-beam lithography in the direction of increased process throughput; elimination of time consuming scanning.
ion-beam lithography

ion reflection an incident ion is reflected from the solid surface instead of being implanted or causing sputtering of the bombarded solid.

ion, single-charged atom ionized to a single charge level, i.e. atom deprived of one electron only; e.g. Ar^{1+}; in conventional discharge in gas single-charge ions dominate population of ions; single-charge ion features very low potential energy and needs to acquire high kinetic energy through acceleration to perform useful function (e.g. sputtering).
ion multicharged

ion sputtering see *ion milling.*

ionic bond a bond in compound materials formed between the electro-negative and electropositive elements; two ions attract each other and are held together by the simple electrostatic force; fairly strong bond; hard substances featuring high melting high and high boiling point; many compounds of interest in semiconductor technology feature combination of covalent and ionic bonds (e.g. GaAs).
covalent bond, metallic bond

ionic conduction electrical conduction in an insulator due to the motion of ions rather than electrons; slow process featuring high activation energy; example: motion of sodium ions (Na^+) in SiO_2.
activation energy, sodium

ionization process in which an atom is converted into an ion typically by giving away one or more of its electrons and assuming a positive charge.
ion, ion single charged, ion multicharged

ionization energy an energy needed to ionize an atom; term commonly refers to the energy needed to ionize (activate) dopant atom in semiconductor, i.e. to have dopant atom to release one free electron (donor) or one free hole (acceptor); dopants featuring low ionization energy are desired.
dopant, dopant activation

IPA see *isopropyl alcohol.*

IPA drying a wafer drying technique using isopropyl alcohol (IPA); wet wafer is exposed to hot vapor of IPA which displaces water from the wafer surface; IPA evaporates during wafer cooling leaving surface moisture-free; the most common wafer drying method in advanced semiconductor manufacturing.
isopropyl alcohol, Marangoni drying, spin drying

IPL see *Ion Projection Lithography.*

IR see *infrared radiation.*

113

iron, Fe common metallic contaminant in semiconductor processing; introduces energy levels in Si bandgap if thermally activated; may originate from several sources in process environment including water and cleaning chemicals as well as corrosion in the stainless steel gas delivery lines.
metallic contaminants

Irvin's curves the curves illustrating changes of the surface dopant concentration as a function of the sheet resistance-junction depth product for different dopant concentration in the bulk of semiconductor substrate at 300 K.
dopant, junction depth, sheet resistance

isolation in semiconductor terminology term typically refers to electrical isolation between two adjacent devices in an integrated circuit, e.g. between *P*-MOSFET and *N*-MOSFET forming CMOS cell; also conducting part of the circuit/device (contacts and interconnects) need to be electrically isolated.
LOCOS, mesa, STI, inter-layer dielectric, spacer

isopropyl alcohol, IPA also known as 2-propanol or isopropanol; colorless, flammable, not safe to drink; highly volatile organic solvent with many industrial and household uses; in semiconductor manufacturing used primarily for wafer drying.
IPA drying

isotropic etching an etch in which rate of etching reaction is the same in any direction, i.e. is non-directional; undesired in the definition of very tight geometrical features, but needed in the case of extensive lateral etching, e.g. in MEMS release.
anisotropic etch, MEMS release

isotype junction the junction involving two semiconductors featuring different chemical composition, but the same conductivity type (*n* or *p*).
heterojunction

ITO see *indium tin oxide*.

J

JBS see *Junction Barrier Controlled Schottky.*

JFET see *Junction Field Effect Transistor.*

JLT see *junctionless transistor.*

Josephson effect a tunneling current flow between two supercon-ductors separated (coupled) by the ultra-thin insulator in the absence of electric filed.

Josephson junction a device designed to implement Josephson effect; made by sandwiching non-superconducting material between two layers of superconducting materials; listed here because of its potential for being used in the very high-speed digital (logic) electronics replacing semiconductor based devices.
Josephson effect

Joule heating the generation of heat in the metal by passing an electric current; also known as resistance heating and ohmic heating; commonly used in the tools employed in semiconductor processing.
furnace, inductive heating

junction in semiconductor materials and devices a region in which two materials featuring different work functions are brought to contact; e.g. *p-n* junction and metal-semiconductor junction (contact).
work function, junction p-n, metal-semiconductor contact

Junction Barrier Controlled Schottky, JBS a Schottky diode based power rectifier with arrays of *p-n* junctions built into the drift region.
Schottky diode

junction breakdown the sudden increase of current flowing across the reverse biased junction; under certain conditions results in the irreversible, permanent damage to the junction and makes it unable to block the current; does not have to be destructive if dissipated power remains within the safe limits.
breakdown, avalanche breakdown, Zener breakdown

junction capacitance a capacitance associated with the variations of the charge in the space charge region of the junction such as p-n junction, metal-semiconductor (Schottky) diode or oxide-semiconductor junction in the MOS structure.
junction, space-charge region

junction depth term typically applies to p-n junction; depth (measured from the surface of semiconductor) of the plane in p-n junction region at which concentration of acceptors is equal to the concentration of donors; determined by the energy of implanted dopant atoms or time of doping by diffusion.
metallurgical junction, doping

Junction Field Effect Transistor, JFET Field Effect Transistor, FET, with a p-n junction acting as a gate; channel's conductivity is controlled by changing width of the space charge region associated with a p-n junction in response to the changing gate voltage, i.e. the voltage biasing a p-n junction gate in reverse.
field effect transistor, MESFET, MOSFET

junction isolation an isolation scheme using reverse biased p-n junction to electrically isolate adjacent transistors in the integrated circuits.
LOCOS, shallow trench isolation

junction, p-n see *p-n junction*.

junction, metal-semiconductor see *metal-semiconductor contact*.

junction spiking see *spiking*.

junction temperature temperature of the junction (p-n, Schottky) during device operation; high junction temperature causes reliability problems; depends on the thermal conductivity (i.e. ability to dissipate heat) of semiconductor; a maximum (critical) temperature at which junction remains operational needs to be established.
critical temperature, junction breakdown

junctionless transistor a MOSFET in which source, channel and drain feature the same conductivity type, i.e. there are no p-n junctions on n^{+}-n

116

junctions on the way of current flow from the source contact to the drain contact.
MOSFET, drain current

K

k Boltzman constant: 8.617342×10^{-5} eVK^{-1}.

k dielectric constant; reflects material ability to store electric charge; unitless; varies from 1 ($k = 1$ for air) to thousands for complex compounds such as titanates and niobates.
high-k dielectric, low-k dielectric

Kaufman source a source of ions commonly used in ion milling equipment.
ion milling

"killer" defect a colloquial term referring to the defect of material or device layout which is a reason for the catastrophic failure of the device; e.g. high density of charge traps in the gate oxide or severe deformation of the interconnect lines.
oxide breakdown, interconnect line

kink site during the epitaxial growth a single atom layer forms an edge with itself and creates a kink site.
epitaxy, MBE

Kirk effect see *base pushout*.

Kirkendall effect concerns exchange of atoms by diffusion at the contact between two different metals; may lead to the malfunctions of the contacts and joints in semiconductor devices.
interdiffusion

KOH potassium hydroxide; used in silicon etching solutions.
CP-4 etch, Secco etch, Sirtle etch

KrF excimer laser an excimer laser featuring 248 nm emission wavelength; used for the photoresist exposure in deep-UV (DUV) photolithography serving 130 nm - 180 nm technology nodes.
excimer laser

Kronig-Penney model a simplified model for an electron in one-dimensional periodic potential; important in derivation of the energy bands and energy band gaps concepts in semiconductors.
energy bands

L

lamp cleaning a brief exposure of semiconductor wafer surface to the halogen-quartz lamp illumination in oxygen containing ambient (including ambient air); increases wafer temperature up to 300 °C and causes volatilization of weakly adsorbed surface organic contaminants; useful in "refreshing" of semiconductor/substrate surfaces in semiconductor manufacturing.
Rapid Optical Surface Treatment, Rapid Thermal Cleaning

lamp heating use of high-power IR lamps for the purpose of increasing wafer temperature; in contrast to other modes of thermal processing (see *furnace*), allows very brief wafer exposure to high temperature (typically not exceeding 1100 °C); low thermal budget process; must be implemented in the way preventing thermal damage of the wafer.
furnace, IR, Rapid Thermal Processing, thermal budget

Langmuir-Blodgett film a film comprised of a single monolayer of an organic material; deposited while slowly withdrawing a solid substrate from the liquid containing organic material; each subsequent immersion/ withdrawal cycle adds a monolayer of organic material to the substrate surface, hence, precise control of the film thickness is possible.
organic semiconductor

LASCR Light-Activated Silicon Controlled Rectifier.

LASER Light Amplification by Stimulated Emission of Radiation; laser; laser is the most common source of coherent, monochromatic radiation; semiconductor lasers are based on *p-n* junction implemented using complex multilayer structures allowing stimulated emission.
semiconductor laser, stimulated emission

laser cleaning use of lasers to remove particles from the wafer surface; particles are dislodged due to the very rapid evaporation of moisture (ablation) from underneath the particle on the wafer surface; features less

than 100% particle removal efficiency, and hence, is not broadly used in commercial applications.
particle

laser diode see *semiconductor laser.*

laser heating allows localized (also in terms of depth) rapid heating of semiconductor wafers and other solids; very useful when local heating is needed, e.g. in local CVD processes such as those used to repair masks; type of laser (wavelength of radiation) is selected depending on the type of material to be heated.
RTP, mask repair, CVD

laser interferometry the endpoint detection method in dry etching tools; completion of the etching process is determined based on the detection of change in the optical characteristics of the laser beam reflected from the etched surface.
endpoint detection

laser, semiconductor see *semiconductor laser.*

latch-up a highly undesired ("parasitic") effect occurring in CMOS devices; condition under which significant current flows through Si substrate between NMOS and PMOS parts of CMOS cell and degrades its performance; it occurs when under certain bias conditions two parasitic bipolar transistors resulting from CMOS configuration "latch" and provide high conductivity path between NMOS and PMOS parts of the cell; various CMOS designs were conceived to prevent latch-up; implementation of CMOS technology on SOI substrates is an ultimate solution to the CMOS latch-up challenge.
CMOS, SOI

lateral diffusion diffusion of dopant atoms in the direction parallel to the surface of semiconductor; undesired in semiconductor device manufacturing as it causes lateral distortion of the device geometry.
diffusion, vertical diffusion

lateral etching an isotropic etching proceeding also in the direction parallel to the surface; undesired, and hence avoided, in the case when very tight geometrical features need to be defined by etching; desired and used in select MEMS etching applications.
isotropic etch, anisotropic etch, MEMS release

lateral transistor a bipolar transistor (BJT), typically *p-n-p*, in which current flows across the base in the direction parallel to the wafer surface; in contrast to conventional vertical bipolar transistor in which current across the base flows in the direction normal to the surface; lateral transistor features inferior characteristics as compared to vertical transistor due to the inferior control over the base width in the lateral configuration.
base width, bipolar transistor

LATID Large-Angle-Tilt Implanted Drain; part of the drain engineering strategy in ultra-small geometry CMOS.
drain engineering, ion implantation

lattice a periodic three dimensional arrangement of atoms in a crystal; specific to a given material; formed by infinite repetition of unit cell in all three spatial directions.
crystal lattice, cubic system, unit cell

lattice cell same as *unit cell*; repeated in all three spatial directions to form a crystal lattice.
lattice, unit cell

lattice constant, *a* distance between atoms in cubic-cell crystals; in non-cubic crystals more than one lattice constants are identified; a measure of structural compatibility between various crystals.
lattice, lattice mismatch, lattice matched structures

lattice matched structure a structure consisting of ultra-thin layers of different single-crystal semiconductors (typically III-V) which are compositionally adjusted such that they feature matching lattice constants; allows changes of energy gap from layer to layer maintaining the same crystallographic structure throughout the entire stack; common in advanced LEDs and laser diodes.
bandgap engineering, lattice constant, superlattice

lattice mismatch a measure of difference (in %) between lattice constants of two different single-crystal semiconductors brought to contact by deposition of one on top of another.
pseudomorphic material

lattice mismatched structure term refers to the situation where two materials featuring different lattice constants are brought together by deposition of one material on top of another; in general, lattice mismatch will prevent growth of defect-free epitaxial film unless an adequate buffer layer is formed or thickness of the deposited film is below certain critical thickness h_c; in this last case lattice mismatch is compensated by the strain in the film.
pseudomorphic material, strained layer superlattice

lattice point a point at which crystal grid lines intersect and which represents the positions occupied by the atoms in a crystal unit cell.
crystal, unit cell

lattice scattering an electrostatic interaction between moving charge carriers and vibrating lattice atoms.
scattering

LCC Leadless Chip Carrier; type of IC package.
package

LCD see *Liquid Crystal Display*.

LCD LED a non-emissive Liquid Crystal Display which uses LEDs as a source of the backlight.
liquid crystal display, light emitting diode, non-emissive display

LDD see *Lightly Doped Drain*.

LDMOSFET Laterally Diffused MOSFET.

leadframe a metal frame that provides external electrical connection to the packaged chip.
package

leak term referring to the malfunction of the vacuum system; a discontinuity of the vacuum enclosure making it impossible to obtain and maintain high vacuum.
vacuum

leak detector apparatus designed to detect leaks in vacuum systems.

leakage current uncontrolled ("parasitic") current flowing across region(s) of semiconductor device in which no current should be flowing; e.g. current flowing across the gate oxide in MOS structure, or excessive (higher than resulting from the physical properties of device) current flowing across reverse biased *p-n* junction, or Shottky diode.
reverse bias

LEC Liquid Encapsulated Czochralski growth; a modified version of CZ process devised to maintain desired composition of the melt by preventing evaporation of its components.
Czochralski growth

LED see *Light Emitting Diode*.

LED display an emissive display; a display composed of the Light Emitting Diodes which convert electric signal directly into light; can be constructed using inorganic or organic semiconductor LEDs
non-emissive display, flexible display, OLED

LED light bulb a light bulb using LEDs as a source of light; the most common light bulb in everyday lighting applications; by far superior to incandescent bulbs in terms of efficiency.
Light Emitting Diode, white LED

LEED Low Energy Electron Diffraction; a method for characterization of physical features (crystallographic structure) of solids.
HEED, RHEED

LER Line-Edge Roughness; term commonly used to describe roughness of the edge of the exposed and developed photoresist.
Photoresist

LET see *Light Emitting Transistor*.

lifetime the period of time between generation/injection of a minority carrier in semiconductor and its annihilation via recombination.
minority carrier, generation, recombination

lift-off a process allowing patterning of thin films without etching; typically used to define geometry of hard to etch metals (e.g. gold); metal

is lifted-off in selected areas by dissolving/ashing underlying photoresist; also used to pattern films composed of nanodots or nanowires.
photoresist, ashing, nanodot, nanowire

light emitter a semiconductor device emitting radiation wavelength (energy) of which is determined by its bandgap; two types of semiconductor emitters: LEDs and lasers; semiconductor light emitters are most commonly configured as diodes.
LASER, LED, LET, radiation wavelength λ

Light Emitting Diode, LED a semiconductor device emitting light; two-terminal rectifying device made using direct bandgap semiconductor compounds; energy generated during recombination processes in the space charge region of the junction (typically *p-n*) is released in the form of light; spontaneous emission as opposed to stimulated emission in lasers; wavelength of emitted radiation depends on the semiconductor's bandgap; LEDs covering radiation spectrum from infrared to violet are available; properly engineered can act as a source of white light; broadly used in lighting and display applications.
EELED, SELED, OLED, white LED

Light Emitting Transistor, LET thin-film transistor made out of inorganic or organic semiconductors capable of emitting light.
inorganic, organic semiconductor

lighting, semiconductor lighting the use of efficient semiconductor LEDs instead of inefficient conventional incandescent (hot filament) or fluorescent (gas-discharge) bulbs in everyday lighting applications; a market dominating lighting technology.
LED, OLED

Light Point Defect, LPD a defect (e.g. particle) revealed on the illuminated surface due to the light scattering; detection of such scattering is a foundation of the particle detection and counting processes in semiconductor technology.
particle counter

lightly doped drain, LDD the reduced doping of the drain region in very small geometry MOS/CMOS transistors; part of the drain engineering strategy in advanced CMOS; designed to control drain-substrate breakdown; reduced doping gradient between the drain and the

channel lowers electric field in the channel in the vicinity of the drain; implementation: moderate implant before spacer formation, heavy implant after spacer formation.
drain engineering, drain extension, spacer

limited-source diffusion also known as a drive-in process; concentration of diffusant (dopant) on the surface decreases during the limited-source diffusion process through dopants redistribution into the substrate in the direction of the concentration gradient; no additional dopant atoms are incorporated into wafer, hence, limited-source.
dopant redistribution, unlimited-source diffusion, drive-in

line defect a dislocation; part of the crystal slips with respect to another part under the stress which is beyond the elastic limit of a given crystal.
crystal defect, dislocation

linearly graded junction a junction in which doping profile depends linearly on the distance from the junction plane.
abrupt junction

liquid crystal, LC a matter in the state which shows characteristics of the conventional liquid, because it flows like liquid, and at the same time characteristics of the solid crystal, because its molecules may be oriented in the crystal-like way; light transmission characteristics of the LC can be altered by the applied voltage which is a working principle of the non-emissive liquid crystal display.
liquid crystal display

liquid crystal display, LCD a non-emissive display composed of the matrix of LC pixels; the image is generate by altering the transmission of externally generated light (thus, non-emissive) through each pixel by applying voltage; controlled by the arrays of thin film transistors (TFT); by definition requires backlighting to operate.
liquid crystal, thin film transistor, LCD LED, AMLCD

liquid phase epitaxy, LPE epitaxial deposition carried out in the gas phase.
epitaxy

lithography process used to transfer pattern to the layer of resist deposited on the surface of the wafer; masked or direct write; type of

lithography depends on the wavelength of radiation used to expose resist: photolithography (or optical lithography) uses UV radiation, X-ray lithography uses X-ray, e-beam lithography uses electron bean, ion beam lithography uses ion beam to expose the resist.
photolithography, e-beam-, ion beam lithography, X-ray lithography

Local Oxidation of Silicon, LOCOS   isolation scheme commonly used in MOS/CMOS silicon IC technology; thick (in the range of 500 nm) pad of thermally grown SiO_2 separates adjacent PMOS and NMOS transistors in the CMOS cell; local oxidation is accomplished by using silicon nitride, Si_3N_4, to block off oxidation of Si in selected areas, hence, "local" oxidation.
channel stop, latch-up, silicon nitride, shallow trench isolation

local planarization   surface planarization process based on the oxide (BPSG) reflow; results in the flat surface over the limited areas on the wafer surface; as opposed to global planarization accomplished by means of CMP.
global planarization, CMP, Boro-Phospho-Silicate Glass

LOCOS   see *Local Oxidation of Silicon.*

logic IC   an integrated circuit performing logic (digital) functions using arrays of logic gates integrated on the semiconductor chip; technology driver in IC engineering; together with memory ICs part of the digital class of integrated circuits.
logic gate, digital IC, microprocessor

logic gate   a basic building block of the logic (digital) circuit; typically features two inputs and one output; there are seven types of logic gates carrying out corresponding logic functions (e.g. AND, NAND, NOR); typically used in combination to perform complex logic functions; arrays of logic gates comprise a logic integrated circuit (IC); state-of-the art logic gates are implemented using CMOS logic.
logic IC

long channel   a channel in the MOSFET long enough to be not affected by the deleterious short channel effects.
channel, short-channel effects

long-range order periodicity of atoms arrangement in crystalline solids; in single-crystal solids long-range order extends over the entire piece of material; in polycrystalline solids the long-range order exists only within limited grains; amorphous solids feature no long-range order.
crystal, amorphous, grain

LOP Low Operating Power; term refers to the logic devices in which a lead concern is to keep bias voltage and current as low as possible to reduce power consumption which is an issue of particular importance in mobile devices.
logic IC, LSTP

low-energy implantation an ion implantation process carried out at low ion energy so that the junction formed is shallow (junction depth is proportional to implant energy).
shallow junction, ion implantation, projected range

low-high, *l-h*, junction a junction in which potential barrier is formed not by bringing to contact semiconductors of opposing conductivity types (*n* and *p*), but heavily doped and lightly doped semiconductor of the same conductivity type.
junction p-n

low-*k* dielectric a dielectric material featuring dielectric constant *k* lower than 3.9 (which is *k* value of SiO_2); used to insulate adjacent metal lines (interlayer dielectric, ILD) in advanced integrated circuits; low *k* reduces undesired capacitive coupling, and hence "cross talk", between lines; a key element of the multilevel metallization scheme.
interconnect, ILD, multilevel metallization

low-pressure mercury (Hg) lamp a source of UV radiation featuring spectrum that includes high intensity lines at 253 nm and 185 nm both needed to generate ozone used in UV/ozone cleaning; can be used in photolithography if higher intensity KrF excimer laser (248 nm wavelength) exposure systems are not available.
ozone, UV/ozone cleaning

low-pressure oxidation a process of thermal oxidation of silicon carried out at the reduced partial pressure of oxidizing species which decreases oxide growth rate at any given temperature.
thermal oxidation, high-pressure oxidation

Low-Temperature Epitaxy, LTP typically below 600 °C; allows formation of very sharply defined layers, but requires extremely careful preparation of the substrate surface to assure growth of the defect-free epitaxial layer.
epitaxy

Low-Temperature Oxide, LTO an oxide deposited by CVD at low temperature (typically below 500 °C); used mostly in BEOL processes as well in the manufacture of devices using semiconductors not resistant to high temperature (e.g. GaAs).
back-end-of-line, CVD

Low-Pressure CVD, LPCVD chemical vapor deposition process carried out at the reduced pressure; improves conformality of coating and purity of the films as compared to atmospheric pressure CVD (APCVD).
conformal coating, CVD, APCVD

LPCVD see *Low-Pressure CVD.*

LPD see *Light Point Defect.*

LPE see *Liquid Phase Epitaxy.*

LSMCD Liquid Source Misted Chemical Deposition; a thin film deposition technique which uses liquid precursors and is based on the mist deposition method.
mist deposition, PLD

LSTP Low Standby Power; term refers to logic devices in which prime concern is to keep standby current as low as possible.
logic IC

LTE see *Low-Temperature Epitaxy.*

LTO see *Low-Temperature Oxide.*

luminescence the emission of light by the material without involving heat; luminescence can be stimulated by the range of interactions other then heating such as illumination or electric field.
photoluminescence, electroluminescence

M

m-s contact   see *metal-semiconductor contact*.

magnetic CZ (Czochralski), MCZ   a single-crystal growth process in which magnetic field is applied to the melt for the purpose of disrupting convection currents in the melt; existence of the convection currents may result in non-uniform incorporation of dopant atoms into the growing crystal as well as non-uniform distribution of alien elements such as oxygen and carbon in the growing crystal; overall, MCZ improves radial uniformity of Czochralski (CZ) grown single-crystals.
Czochralski crystal growth

magnetic (ferromagnetic) semiconductor   a semiconductor material (e.g. GaAs) which is converted into ferromagnetic material by adding magnetic dopants (e.g. manganese); magnetic semiconductors are the basis of spintronics technology.
spintronics

magnetically confined plasma   a plasma in which magnetic field is used to confine electrons in the plasma and by doing so to increase ionization efficiency, and hence, to form a high-density plasma.
high-density plasma

Magnetically Enhanced RIE, MERIE   a reactive ion etching process in which plasma is confined by magnetic field; what results is an increase of the ionization efficiency producing denser plasma, increased etch rate and reduced surface damage.
high density plasma, magnetically confined plasma, reactive ion etching

magnetoresistance   term refers to the changes of the electrical resistance of solids (including some semiconductors) under the influence of magnetic field.

magnetron   a process tool designed to implement magnetron sputtering.

magnetron sputtering   sputtering process in which plasma is confined by the magnetic field; ionization efficiency is increased leading to higher density of ions which in turn increases sputtering rate.
sputtering, magnetically confined plasma

majority carriers charge carriers (electrons or holes) which dominates conductivity of a given semiconductor; one of two carrier types featuring equilibrium concentration higher than that of the other type; electrons are majority carriers in n-type semiconductor ($n_0 \gg p_0$) while holes in p-type semiconductor ($p_0 \gg n_0$).
minority carriers

manufacturing yield see *yield*.

Marangoni drying a wafer drying technique; wafer is gradually withdrawn from rinsing water into the vapor of isopropyl alcohol (IPA) and nitrogen; due to the gradient of surface tension, water from the surface of the wafer is pulled back into the body of water leaving emerging surface water-free.
IPA drying, rotagoni drying

mask an object used to define desired geometries on the surface of the wafer by mechanically (mechanical mask) or optically (photomask mask) blocking access of physical species (e.g. during evaporation) or short-wavelength radiation to the selected parts on the wafer surface; alternatively, a desired pattern can be reflected off the mask and projected onto the surface of the wafer (reflective mask); depending on the lithography process transparent and reflective masks are distinguished.
photomask, mask: -mechanical, - reflective

mask, Extreme UU Lithography see *reflective mask*.

mask making a process of manufacturing masks; in the case of photomasks based on the direct write e-beam lithography.
e-beam lithography, mask

mask, mechanical also known as a shadow mask; a thin sheet of metal or plastic with properly shaped openings cut into it acting as a stencil; used to define crude geometries on the surface of the wafer during physical vapor deposition (evaporation, sputtering) or physical liquid deposition (mist deposition); most commonly used to form metal contacts in crude test devices.
physical vapor deposition, physical liquid deposition

mask, photolithography a mask used in photolithography to block photoresist exposure to UV light in selected areas leaving other areas transparent to UV light; transmission mask; consists of thin layer of chrome or emulsion opaque areas supported by the high quality quartz or glass plate transparent to UV light and referred to as a blank.
blank, chrome mask, photoemulsion mask

mask, reflective a mask used in EUV lithography and X-ray lithography where the wavelength of radiation used for the resist exposure is too short for the conventional transmission mask to accurately reproduce desired patterns (because of the excessive diffraction at the mask edges and absorption in the blank); pattern formed on the surface of the mask is reflected and projected on the surface of the resist covered wafer.
transmission mask, Extreme UV lithography, X-ray lithography

mask repair a process allowing restoration of the original pattern on the surface of the damaged (e.g. scratched) mask by means of CVD combined with local surface heating.
CVD, laser heating

mask, transmission a mask in which desired pattern to be transferred to the layer of resist is composed of regions transparent and regions non-transparent (opaque) to the radiation used for resist exposure; see *mask, photolithography* and *mask, X-Ray lithography*.
mask, reflective mask

mask, X-ray lithography both reflective masks and transmission masks can be used; in the latter case gold used as opaque material is supported by a thin membrane made of material transparent to X-rays of the given wavelength, e.g. Si_3N_4, SiC and others.
mask, transmission masks, reflective masks

masked lithography involves lithography techniques in which desired pattern is established using masks either transmission or reflective; an alternative is a direct write lithography.
mask, mask, transmission, mask, reflective; direct write lithography, photolithography

mass action law in the semiconductor in thermal equilibrium and in the absence of an electric field or illumination the following relationship concerning concentration of electrons and holes holds: $np = n_i^2$ where n

and p is concentration of electrons and holes respectively and n_i is an intrinsic carrier concentration.
intrinsic concentration

mass flow controller, MFC an instrument installed in the gas delivery lines in semiconductor equipment to control the amount of gas delivered to the process chamber; the flow is typically measured in sccm.
sccm

Maxwell-Boltzman statistics describes energy distribution within the group of particles; the number of particles with given $N(E)$ is proportional to $exp(-E/kT)$ where k is the Boltzman constant; does not impose restriction on the number of particles that can have the same energy state; it works for electron energy distribution in a conduction band of non-degenerated semiconductor.
Fermi-Dirac distribution function, Boltzman constant, non-degenerate semiconductor

Maxwell-Boltzmann distribution see *Maxwell-Boltzman statistics*.

Maxwell's equations a set of partial differential equations which describe interactions of electric and magnetic fields and which are at the foundation of electronics, optics and electromagnetics.

MBE see *Molecular Beam Epitaxy*.

MCP Multi-Chip package; a type of package in 3D chip stacking technology.
three-dimensional integration

MCM see *multichip module*.

MCZ see *magnetic CZ*.

mean free path an average length covered by the charge carrier in the solid's lattice between two successive collisions; the term applies also to the distance between collisions covered by the species in the gas-phase.
scattering

mechanical mask see *mask, mechanical*.

mechanical pump in semiconductor terminology synonymous with a mechanical vacuum pump; operation is based on the mechanical action, e.g. rotation, causing displacement of gas; effective in evacuating air/gas down to about 10^{-4} torr; used as a roughing pump in vacuum equipment.
roughing pump, Roots pump, vacuum pump, torr

megasonic agitation a megasonic energy at the frequency in the range from ~ 500 kHz - ~ 1 MHz range is applied to the liquid to create sonic waves (sonic pressure) to enhance cleaning/rinsing action; more effective and less potentially damaging to the wafer (reduced cavitation) than ultrasonic agitation; standard feature in the immersion cleaning tools.
immersion cleaning, wet bench

megasonic cleaning a wafer cleaning process which uses sonic waves (megasonic agitation) generated in cleaning solution to increase efficiency of particle removal process; most commonly carried out in conjunction with APM cleans.
megasonic agitation, APM, particle removal, wet bench

megasonic scrubbing essentially synonymous with megasonic cleaning except that the term "scrubbing" implies "heavy duty" cleaning such as cleaning after CMP aimed at the removal of polishing slurry from the wafer surface.
post-CMP cleaning, slurry

memory cell a semiconductor structure electrical state of which (e.g. conductive, non-conductive) can be altered and then retained in the altered state; in this way a bit of information can be stored.
RAM, ROM

memory IC semiconductor integrated circuit designed to store data either temporarily by the random access memory (RAM), or permanently by the read only memory (ROM); together with logic ICs comprises a digital class of integrated circuits.
DRAM, SRAM, logic IC, digital IC

MEMS Micro-Electro-Mechanical Systems; micro-machined in silicon using the same techniques as those used in the manufacture of electronic silicon devices; typically integrated on the same chip with electronic microcircuits; take advantage of the excellent mechanical properties of

silicon; generally fall into two categories of microsensors and micro-actuators; depending on application operation based on electrostriction, or electromagnetic, thermoelastic, piezoelectric, or piezoresistive effects.
NEMS, Deep Reactive Ion Etching

MEMS release etching of sacrificial material, typically oxide, from the inside of the MEMS structure; process releases moving parts which were originally supported by the sacrificial oxide.
sacrificial oxide, isotropic etch, lateral etch, anhydrous HF

mercury probe a probe with a mercury tip allowing temporary contact to the semiconductor surface; used to measure electrical characteristics of semiconductors without formation of the permanent metal contact on the surface.

MERIE see *Magnetically Enhanced RIE.*

mesa isolation in the circuits formed on semi-insulating substrates the adjacent devices are isolated by etching through conducting layer and creating "islands" in which devices are formed.
isolation

mesa transistor term refers to the way early bipolar transistors featuring diffused base and emitter were configured; eventually replaced by planar transistor configuration.
planar transistor

MESC port Modular Equipment Standardization Committee; a standardized port; developed in order to facilitate mechanical integration of process modules in cluster tools and minienvironments used in semiconductor manufacturing.
cluster, minienvironment, integrated processing

MESFET see *Metal-Semiconductor Field Effect Transistor.*

metal a material, typically solid, rarely liquid, inherently featuring very high electrical conductivity; in metals atoms are held together by the force of metallic bond; in the energy band structure of metals the conduction and valence bands overlap, and hence, there is no energy gap; upper most energy band is only partially filled with electrons;

electrons can move around almost freely under the applied voltage; indispensible in semiconductor device/circuit engineering for contacts and interconnects.
contact, interconnect, semiconductor, insulator

metal contact in semiconductor terminology synonymous with *metal-semiconductor contact.*

metal MOS gate metals are not used as gate contact material in MOS/CMOS devices in conjunction with SiO_2 and nitrided gate SiO_2; instead, conducting poly-Si is used due to the work function matching work function of Si substrate (requirement for the low threshold voltage of a MOSFET); metals as MOS gate contacts are used in cutting-edge CMOS technology in conjunction with high-*k* dielectrics (HKMG) replacing poly-Si gate contacts; poly-Si forms an SiO_x layer at the interface with gate dielectrics (see *poly depletion*) while metal gates do not.
high-k dielectrics, MOS gate, poly depletion, HKMG

Metal-Organic Chemical Vapor Deposition, MOCVD a CVD process which uses metal-organic compounds as the source materials; metal-organics thermally decompose at temperatures lower than other metal containing compounds; method used in epitaxial growth of very thin films of III-V semiconductors and in other deposition processes.
Chemical Vapor Deposition, epitaxy

Metal-Oxide-Semiconductor, MOS a three-layer structure (typically M-SiO_2-Si) in which concentration of charge carriers in semiconductor's sub-surface region is controlled by potential applied to the metal contact, or in other words, by the field effect; MOS gate can invert sub-surface region of semiconductor underneath metal-oxide gate; it works only if no excessive leakage current flows across the oxide; core of the MOS Field Effect Transistors, and hence, CMOS.
accumulation, depletion, inversion, field effect, MOSFET

Metal-Oxide Semiconductor Field Effect Transistor; MOSFET a Field-Effect Transistor, FET, with MOS structure acting as a gate; current flows in the channel between source and drain; channel is created by applying adequate potential to the gate contact and inverting semiconductor surface underneath the gate; MOSFET structure is

implemented almost uniquely with Si and SiO_2 or high-k dielectric acting as a gate oxide; efficient switching device which dominates logic and memory applications; PMOSFET (p-channel, n-type Si substrate) and NMOSFET (n-channel, p-type Si substrate) combined form basic CMOS cell.
CMOS, Field-Effect Transistor, JFET, MESFET

metal-semiconductor contact a key component of any semiconductor device; depending on materials involved in the contact its properties differ drastically; ideally, contact is ohmic (linear, symmetric current-voltage characteristic, contact resistance is negligible) when the work function of metal matches the work function of semiconductor (no potential barrier at the interface); contact is rectifying (non-linear, highly asymmetric, diode-like current-voltage characteristic) when the work functions of metal and semiconductor are different (potential barrier at the interface results); commonly referred to as a Schottky diode.
ohmic contact, Schottky diode, contact resistance

Metal-Semiconductor Field Effect Transistor; MESFET a Field Effect Transistor, FET, with metal-semiconductor contact (Schottky diode) acting as a gate; channel's conductivity is controlled by the gate voltage which changes the width of the space charge region associated with metal-semiconductor contact; allows implementation of field effect transistor with semiconductors which do not have high quality native oxide (e.g. GaAs), and hence, are not compatible with MOS gate approach.
metal-semiconductor contact, space charge region, JFET, MOSFET

metallic bond positive metal ions are embedded in an electron gas that permeates the entire solid; the material is held together by electrostatic interaction between the ions and electrons; electrons can move freely throughout the material without significant change in their energy; materials featuring metallic bond are excellent conductors.
covalent bond, ionic bond

metallic contaminants atoms of alien metals inadvertently deposited on the semiconductor surface during device processing; common metallic contaminants: Fe, Al, Cu, Ca, Na; originate from process chemicals, water and process tools; cause major reliability problems when activated during elevated temperature processes; in the case of Si

surface concentration of metallic contaminants should not exceed 10^9 cm^{-2}; designated cleans are applied to remove metallic contaminants from the surface.
HPM, RCA clean

metallization   a process forming metal contacts and interconnects in the manufacturing of semiconductor devices.
chemical vapor deposition, physical vapor deposition

metallurgical junction   term refers to *p-n* junction; an interface between *n*-doped and *p*-doped regions of the junction, the plane in which concentration of acceptors equals concentration of donors ($N_A = N_D$).
junction depth

metamaterials   a man-made materials engineered to display properties not observed in nature; as an example a semiconductor metamaterials may display negative refractive index; synthetized to display desired characteristics using multiple components configured from the conventional materials such as metals or plastics.
refractive index

MFC   see *mass flow controller*.

MFMISFET   Metal-Ferroelectric-Metal-Insulator-Semiconductor Field Effect Transistor.

MGHK   see *HKMG*.

Micro-Electro-Mechanical System   see *MEMS*.

microloading   the effect minimizing etch rate of very high-aspect ratio geometrical features at the reduced pressure.
etching, high-density plasma, aspect ratio

micrometer, μm   one millionth of a meter, 10^{-6} m; 1 μm = 1,000 nm; terms microelectronics, micromachining, microtechnology, micro-fabrication, etc. refer to this unit of length; also referred to as a "micron".
nanometer

microprocessor central processing unit (CPU) fabricated on one or more chips; contains basic elements of a computer, including logic and control, that are needed to process data; essentially a logic chip.
logic IC

microwaves an electromagnetic radiation with wavelengths ranging from 10^{-3} m to 1 m and corresponding frequencies of 300 MHz to 300 GHz; partially overlaps with a broad range of radio frequencies (RF).
Radio Frequency

microwave plasma, MW plasma plasma generated using microwave frequency signal; typically 2.45 GHz; generates denser plasma than more conventional RF (13.56 MHz) signal.
high density plasma, RF plasma

mil one thousandth of an inch, or 25.4 μm.

Miller capacitance result of the *Miller effect*; in the case of the MOSFET a parasitic capacitance between the gate and the drain; has an adverse effect on the transistor performance at the very high frequencies.
Miller effect

Miller effect in general, an increase in the equivalent input capacitance of an amplifier due to the effect of increased capacitances between the input and output terminals.

Miller indices a set of coordinates defining orientation of the specific crystallographic plane in the crystal; a combination of three digits, either 1 or 0, e.g. (100), (111), etc. determines how the plane intersects the main crystallographic axes of the crystal; commonly used to define orientation of the crystals surface.
surface orientation, crystal planes

MIM Metal-Insulator-Metal; a capacitor structure.

MIMIC, MMIC Microwave Monolithic Integrated Circuit; an IC operating in the microwaves regime; typically fabricated using semiconductors featuring high electron mobility, e.g. GaAs.
Microwaves

minienvironment an ultra-clean environment created within the limited space surrounding individual process tool and not in the entire cleanroom; sealed off SMIF boxes are used to move wafers from one tool/minienvironment to another; more economical approach to semiconductor manufacturing than "ballroom" approach in which ultra-clean environment is created in the entire room.
ballroom cleanroom, SMIF

minority carrier one of two carrier types (electrons of holes) featuring equilibrium concentration lower than that of the other type; holes in *n*-type semiconductors, electrons in *p*-type semiconductors.
majority carriers

minority carrier lifetime time between generation and recombination of the minority carrier; a measure of the structural integrity of semiconductor material; short m.c.l. indicates high density of defects acting as the carrier recombination sites.
lifetime, generation, carrier injection, carrier recombination

MIS Metal-Insulator-Semiconductor structure; term more general than MOS (metal-oxide-semiconductor); applicable to the cases where insulator is not an oxide, e.g. Si_3N_4.

MISFET Metal-Insulator-Semiconductor Field Effect Transistor; more general term than Metal-Oxide-Semiconductor FET, but describing the same device structure.
MIS, MOSFET

misfit dislocation crystal defect occurring in heterostructures involving single-crystal semiconductors with mismatched lattice constants; result of the release of stress accumulated in mismatched crystals.
crystal defects, dislocation, lattice mismatch, lattice constant

mismatch in semiconductor terminology synonymous with lattice mismatch.
lattice mismatch

mismatch accommodating layer a layer between two lattice mismatched crystals; provides for a defect-free and strain-free transition between crystals featuring different lattice constants.
buffer layer

mist deposition a deposition technique in which liquid precursor is delivered to the substrate in the form of a very fine mist (submicron droplets); in semiconductor manufacturing can be used to deposit very thin-films of photoresist and dielectrics; also effective in thin film deposition using colloidal solution precursors; best suited for deposition of films thinner than 100 nm.
Physical Liquid Deposition, conformal coating, colloidal solution

mixed signal IC an IC chip incorporating both digital and analog functions.
analog IC, digital IC

MNOS Metal-Nitride-Oxide-Semiconductor; basically a MOSFET in which gate dielectric comprises of two layers, SiO_2 and Si_3N_4; of interest because charge trapping at the SiO_2/Si_3N_4 interface allows non-volatile memory applications of the device.
nonvolatile memory

mobile charge electrically active specie which can move in the MOS gate oxide under the influence of an electric filed; causes severe instabilities of the MOSFETs characteristics, e.g. fluctuations of the threshold voltage V_T; sodium ions Na^+ are the most common mobile charges in SiO_2.
gate oxide, threshold voltage

mobile ion term refers to the ion which can move in the solid in the presence of an electric field.
mobile charge

mobility proportionality factor between semiconductor conductivity and concentration of free charge carriers (electrons and holes); symbol μ; key parameter defining transport of free charge carriers in semiconductor, and hence, semiconductor's conductivity; different for different semiconductors due to the differences in the effective mass of electrons and holes in different semiconductors; higher for electrons than holes due to lower effective mass of the former; strongly depends on carrier scattering.
electron mobility, hole mobility, scattering

MOCVD see *Metal-Organic Chemical Vapor Deposition.*

MODFET Modulation Doping Field Effect Transistor; a type of the heterojunction FET; built on the modulation doping principle; capable of operating at the very low temperatures; typically formed using combination of AlGaAs and GaAs epitaxial layers.
HFET, modulation doping

modulation doping doping of the heterostructure (e.g. AlGaAs-GaAs) implemented in such way that the resulting free electrons are spatially separated from the positive donor ions; as a result, scattering of the moving electrons on the dopant atoms is avoided; also, due to the separation, electrons remain free and mobile even at the very low temperature; MODFET is a transistor built on the modulation doping principle.
MODFET, scattering

MOEMS Micro-Opto-Electro-Mechanical Systems; basically MEMS devices which can control, or can be controlled, by light; based on optical phenomena such as reflection, diffraction, or refraction.
MEMS

Molecular Beam Epitaxy, MBE a process of physical epitaxial deposition (basically a PVD process) carried out in ultra-high vacuum (below 10^{-8} torr) and at the substrate temperature typically not exceeding 800 °C; due to unobstructed (molecular) flow of species to be deposited and chemical cleanliness of the substrate surface a highly controlled growth of ultra-thin epitaxial layers is possible; the highest precision (essentially atomic layer-by-layer) epitaxial deposition method used in semiconductor processing involving in particular III-V compounds; used in fabrication of superlattices, strained layers, quantum-wells, etc.
PVD, superlattice, quantum well, epitaxy by CVD

molecular bond weak bond occurring in inert gases and some organic molecules; a result of the dipolar forces between bonded species.
covalent bond, ionic bond, metallic bond

molelectronics molecular electronics; systems in which electronic functions (normally performed by transistors) are performed by organic molecules.

molybdenum disulfide, MoS₂ transition metal dichalcogenide is a semiconductor featuring indirect bandgap $E_g = 1.23$ eV; a compound

which can be processed into a truly 2D material (single- or few-layers) displaying a direct bandgap $E_g = 1.8$ eV and featuring a range of outstanding characteristics; unlike graphene, it has a bandgap which makes it attractive in digital transistor applications, particularly high-temperature (above 200 °C).
graphene, silicone, phosphorene

molybdenum silicide, MoSi$_2$ contact (ohmic) material in Si technology; resistivity 90-100 μΩ-cm; formed at the sintering temperature of 1100 °C.
silicide, sintering temperature

monolithic IC the entire electronic circuit is built into a single piece of semiconductor (chip); physical properties of semiconductor to a large degree determine performance of the circuit; common integrated circuits such as microprocessors, memories, etc., are all monolithic.
hybrid IC

Moore's law an observation, originated around 1970 and confirmed over several technology generations which states that the number of transistors in cutting edge integrated circuits, and hence, their functional capabilities, double approximately every 2 years.

MOS see *Metal-Oxide-Semiconductor*.

MOS capacitor also referred to as a MOS cap; a metal-oxide-semiconductor structure forming a parallel plate capacitor; a building block of the MOSFET; commonly used as a test structure allowing determination of electrical properties of the materials involved, and hence, predicting their performance in the MOSFET configuration.
MOSFET, C-V characterization, I-V characterization

MOS gate a metal-oxide-semiconductor structure used as a gate which controls output current in the MOSFET; also referred to as a MOS gate stack.
gate stack, HKMG, metal MOS gate

MOSFET see *Metal-Oxide-Semiconductor Field Effect Transistor*.

MOSFET scaling a reduction of the MOSFET's geometry for the purpose of improving transistors performance; the goal is to reduce gate

(channel) length and in the process to maintain device fully operational; scaling rules must be followed to avoid undesired short-channel effects.
scaling rules, short-channel effects, gate scaling

MOST   Metal-Oxide-Semiconductor Transistor; same as *MOSFET*.

MPGA   Metal Pin Grid Array; type of IC package.
Pin Grid Array

MPS   Merged P-I-N/Schottky; type of high power rectifier combining *p-i-n* junction and Schottky diode.
Schottky diode, p-i-n junction

MRAM   Magnetoresistive Random Access Memory; a nonvolatile memory; uses electron spin instead of electric charge to store data; promising candidate for DRAM applications.
nonvolatile memory; DRAM

MTBF   Mean Time Between Failures; a measure of the reliability of process tools used in semiconductor manufacturing.

MTL   Merged Transistor Logic; same as integrated injection logic (I^2L).
integrated injection logic

MuGFET   Multi-Gate FET; a family of MOSFETs in which, in order to increase the gate area without increasing chip area occupied by the transistor the gate is accessed from more than one side; e.g. tri-gate FET, FinFET, gate-all-around FET.
MOS gate, gate all-around, FinFET, tri-gate FET, MOSFET

multi-gate MOSFET   see *MuGFET*.

multi-walled nanotube, MWNT   a coaxial assembly of single-walled nanotubes; in terms of internal structure similar to a coaxial cable; the term applies primarily to carbon nanotubes.
nanotube, single-walled nanotube, carbon nanotube

multicharged ion   see *ion, multicharged*.

Multichip Module, MCM   package containing more than one IC chip.
single chip module

multicrystalline material just like a polycrystalline the m-c material maintains long-range order only within limited in volume grains; it differs from polycrystalline material in that the grains in m-c material are larger and are typically significantly expanded along the direction of crystal solidification; the m-c Si is broadly used in solar cell manufacturing where it offers advantageous cost *vs.* efficiency relationship.
polycrystalline material, single-crystal, solar cell, efficiency

multilayer metallization metal lines or contacts consisting of two or more different metals deposited on top of each other and each serving different purpose; for instance, a layer of aluminum will assure low resistivity of the contact while underlying titanium nitride will act as a barrier preventing aluminum spiking.
barrier metal, spiking

multilayer resist in order to improve resolution of the pattern transfer process more than one layer of resist is used; in the simple version of such scheme a bottom (thicker) layer planarizes the surface while the top (very thin) layer acts as an imaging resist.
imaging resist, planarization

multilevel interconnects, MLI the interconnect scheme in integrated circuits implemented in several levels; metal lines stacked into several levels (more than ten in the most advanced circuits) are electrically isolated from each other by the interlayer low-k dielectric and vertically interconnected through the vias; MLIs are needed in very high density integrated circuits to save space on the surface of the chip, shorten the length of the lines, to minimize *RC* losses, and to prevent current density in metal lines to increase above certain critical value.
interconnect, via, interlayer dielectric, low-k dielectric

multilevel metallization see *multilevel interconnect.*

multiple patterning an approach to pattern definition employed in photolithography in the manufacture of 32 nm technology generation ICs and below; allows an increase of feature density without having to employ shorter exposure wavelength; double patterning is the most common version of m.p.; in conjunction with the immersion photolithography extended applicability of 193 nm exposure tools.
immersion photolithography, enhancement techniques

multiplication factor, *M* defines increase of the reversed biased *p-n* junction current under the avalanche multiplication condition.
avalanche multiplication

N

n concentration of electrons in semiconductor (cm^{-3}); also depiction of semiconductor in which electrons are acting as majority carriers.
p-n junction, majority carriers

n see *ideality factor.*

n see *refractive index.*

***n*-channel MOSFET** see *NMOSFET.*

***n*-type conductivity** electrical conductivity of semiconductor dominated by the electrons acting as charge carriers.
p-type conductivity

***n*-type dopant** a dopant (alien element, donor) introduced into semiconductor material to render it *n*-type.
donor, on implantation, diffusion

***n*-type semiconductor** a semiconductor material in which concentration of electrons is much higher than the concentration of holes ($n \gg p$); electrons are majority carriers and dominate conductivity.
intrinsic semiconductor, p-type semiconductor

NAA Neutron Activation Analysis; a method used to determine chemical composition of materials.

nanocrystalline quantum dot, NQD a zero-dimensional (zeroD) semiconductor nanocrystal smaller than the diameter of the Bohr exciton in the bulk of the crystalline material of the same chemical composition; typically between 2 nm and 10 nm in diameter; due to the quantum confinement the NQDs feature much different electronic and photonic properties than the bulk material of the same chemical composition; physical properties of NQD depend on its size; promising is several both

electronic and photonic applications; often referred to as *nandot* or *quantum dot*.
Bohr exciton radius

nanodot a broad term encompassing crystalline and non-crystalline zero-dimensional (zeroD) material systems of various chemical compositions; in practice "zeroD" indicates nanodot's diameter of less than 10 nm.
nanocrystalline quantum dot, nanowire

nanoglass™ a porous silica (glass); porous SiO_2; porosity (air gaps in *nm* scale) are created in the material in order to reduce its dielectric constant *k* from *k* = 3.9 for non-porous silica to a low as *k* = 1.3 for porous silica.
ILD, low-k dielectric

nanoheteroepitaxy, NHE the process which uses an array of nanoscale pillars sandwiched between the substrate and the epitaxial layer to absorb energy of strain in the lattice-mismatched systems; e.g. potentially useful in depositing GaAs and GaN on Si substrates.
heteroepitaxy, lattice mismatch

nanometer, nm one billionth of a meter, 10^{-9} m [nm]; terms such as nanoelectronics, nanomachining, nanotechnology, nanofabrication refer to this size scale; 1 nm = 10 Å = 0.001 of μm.
angstrom, micrometer, nanotechnology

nanoporous silica see *nanoglass*.

nanoprinting see *soft lithography*.

nanotechnology scientific and technical endeavor in which solid matter is manipulated in the atomic and molecular scale; in semiconductor domain term "nanotechnology" refers primarily to the fabrication of functional, information processing integrated circuits in the nanometer length scale; also applies to the use of nanoscale material systems (nanodots, nanotubes, nanowires) in the fabrication of electronic and photonic devices.
nanometer, gate length, nanodot, nanotube, nanowire

nanotube a tubular molecule; 1D material system commonly used in reference to carbon nanotubes.
carbon nanotube, single-walled nanotube, multi-walled nanotube

nanowire 1D material system; strongly two-dimensionally confined (ultra-small diameter in deep sub-100 nm regime) pieces of a solid in the form of a wire or similar elongated structures; surface to volume ratio of a nanowire is larger than the same ratio of geometrical features that can be define by lithographic means; semiconductor nanowires (e.g. silicon) and metal nanowires (e.g. gold) are of interest.
silicon nanowire, carbon nanotube

native oxide an own oxide of the solid; e.g. SiO_2 is a native oxide of Si, Al_2O_3 is a native oxide of Al, etc.; in the case of Si, spontaneously grows to the thickness of about 1 nm; among semiconductors only in the case of Si and SiC (silicon carbide) native oxides features device-grade properties.
silicon dioxide, silicon carbide

NBTI see *Negative Bias Temperature Instability*.

Negative Bias Temperature Instability, NBTI represents fluctuations of electrical characteristics of the silicon MOS devices when negative bias is applied to the gate at the increased temperature.
bias-temperature stress, MOS gate

negative charge dielectric a dielectric material used in the form of the thin-film in semiconductor technology which features defects that are inherently predominantly negatively charged; e.g. Al_2O_3.
positive charge dielectric

negative resist a resist (e.g. photoresist) which is initially soluble in the developer, but becomes insoluble after irradiation; crosslinking between polymeric chains takes place during irradiation; features lower resolution than positive resist because its volume slightly increases as a result of crosslinking, but is more sensitive.
developer, positive resist, crosslinking

negative resistance term refers to a dynamic resistance; situation in which device current is decreasing with increasing voltage (e.g. in tunnel diodes); observed on the current-voltage characteristics of the device.
tunnel diode, current-voltage measurements

146

NEMS Nano-Electrical Mechanical System; conceptually the same as MEMS, but implemented using nanometer-scale geometrical features; potentially of great importance in the range of applications beyond typical applications of MEMS.
MEMS

Next Generation Lithography, NGL in common usage the term refers to the lithography technique(s) expected to carry the load of pattern definition in 5 nm and below technology generations.
technology generation

NGL see *Next Generation Lithography.*

nickel silicide, NiSi a contact (ohmic) material in Si technology; resistivity 14-20 $\mu\Omega$-cm; formed at the sintering temperature of 400 °C - 600 °C.
silicide, ohmic contact, sintering temperature

nitric oxide, NO a gas used in semiconductor processing as a source of atomic nitrogen or as an oxidant of silicon; highly toxic.
nitrodation, nitrous oxide, nitrided oxide, ONO

nitridation a high-temperature anneal in nitrogen containing ambient (NO, N_2O or NH_3); typically refers to the process during which nitrogen is added to ultra-thin SiO_2 on Si surface.
nitrided oxide

nitrided oxide a nitrogen rich silicon dioxide; formed by exposing SiO_2 to nitrogen containing ambient at high temperature or to nitrogen containing plasma; also a result of thermal oxidation of silicon in NO or N_2O.
nitridation, ONO, nitric oxide, nitrous oxide

nitrides groups III-V semiconductor compounds of nitrogen (N, group V) with elements from the group III, e.g. GaN, BN.
antimonides, arsenides, phosphides

nitrogen, N_2 an inert gas commonly used in semiconductor processing as a carrier gas or an inert ambient; not as chemically inert as argon, but much less expensive; due to the strength of N-N bond in N_2 molecule not used as a source of atomic nitrogen N.
ammonia, nitric oxide, nitrous oxide

nitrous oxide, N₂O a nonflammable gas used in semiconductor processing as a source of atomic nitrogen or as oxidant in thermal oxidation of silicon; nontoxic, known as a "laughing gas".
nitridation, nitric oxide, nitrided oxide, ONO

NMOSFET *n*-Channel MOSFET; a Metal-Oxide-Semiconductor Field Effect Transistor with *n*-type channel, i.e. device in which electrons are responsible for conduction in the channel; built into *p*-type substrate with n^{++} doped source and drain.
channel, PMOSFET

noise in semiconductor engineering a random fluctuations in an electrical signal generated by semiconductor devices; various types of noise are distinguished based on their sources.
flicker noise, thermal noise, telegraph noise

non-contact electrical characterization characterization methods allowing determination of selected electrical characteristics of semiconductor and/or its surface without making a physical contact to the surface; to perform a non-contact measurement the equilibrium of the near-surface region of semiconductor must be disturbed, typically by illumination with light with energy $hv > E_g$ (bandgap).
surface photovoltage, surface charge analysis

non-degenerated semiconductor semiconductor in which Fermi level is located within its bandgap not closer to one of the band edges (either conduction or valance) than $2\ kT/q$.
energy bands, Fermi level, degenerate semiconductor

non-emissive display a display in which image is created by exploiting optical effects induced in the medium through which externally generated light is passing; LCD display is a prime example on a non-emissive display; requires backlight.
LCD display, emissive display

non-radiative recombination a recombination of electrons and holes in semiconductor which does not produce photons (electromagnetic radiation), but mostly phonons; i.e. energy released during the recombination process is in the form other than electromagnetic radiation; occurs in semiconductors featuring indirect bandgap.
phonon, photon, recombination, radiative recombination

non-selective etching the etching process in which etching medium interacts with materials on the wafer surface regardless of their chemical composition; in other words the etch rates of different materials exposed to etch chemistry do not differ significantly.
selective etching, physical etch

non-volatile memory a memory in which information stored is maintained after the power supply is turned off.
volatile memory

normally "off" MOSFET see *enhancement mode MOSFET*.

normally "on" MOSFET see *depletion mode MOSFET*.

np **product** the product of electron (n) and hole (p) concentrations in a given semiconductor in thermal equilibrium is constant and equal to n_i^2.
intrinsic concentration n_i, thermal equilibrium

n-p-n **transistor** a bipolar transistor with n-type emitter and collector and p-type base; displays superior characteristics as compared to its p-n-p counterpart due to the higher mobility electrons rather lower mobility holes acting as minority carriers in the base.
p-n-p transistor, bipolar transistor

NTD Neutron Transmutation Doping.

NSOM Near Field Scanning Optical Microscopy, also known as *SNOM*; optical microscopy based technique; allows characterization of soild surface with nanometer resolution.

nucleation in semiconductor terminology the term usually refers to the first stage of the thin-film formation on solid surfaces; particularly important in the early stage of epitaxial deposition.
thin film, epitaxy

numerical aperture, NA parameter defining geometry of the objective lens used in projection printing in photolithography; determines ability of the lens to collect light diffracted from the mask/reticle.
projection printing, resolution, stepper

149

O

OBIC   Optical Beam Induced Current, a semiconductor material characterization technique based on the analysis of the laser beam induced current.

OEM   Original Equipment Manufacturer.

OFET   Organic Field Effect Transistor; a thin-film transistor (TFT) formed using organic semiconductors.
Field-Effect Transistor, organic semiconductor, thin-film transistor

ohmic contact   a metal-semiconductor contact featuring very low resistance which is independent of applied voltage (may be represented by constant resistance); to form an "ohmic" contact the metal and semiconductor must be selected such that there is no potential barrier formed at the interface or the potential barrier is so thin that charge carriers can readily tunnel through it.
metal-semiconductor contact, tunneling

ohmic heating   see *Joule heating*.

OLED   see *organic LED*.

OLED display   an emissive display fabricated using organic light-emitting diodes (OLEDs); can be flexible if formed on the flexible substrate.
emissive display, flexible display

omega gate   3D MOS gate which shape in cross-section resembles Greek letter "omega" (Ω); variation of FinFET technology.
FinFET, gate allaround, pi-gate

OMVPE   Organo-Metallic Vapor Phase Epitaxy.
epitaxy

on resistance   the resistance between source and drain of the power MOSFET.
power device

one-dimensional (1D) material   see *nanowire, nanotube*.

ONO Oxide-Nitride-Oxynitride; a tri-layer MOS gate dielectric; main function of the nitride phase in the gate oxide is to block diffusion of boron from p^{++} poly-Si gate contact into gate oxide.
boron penetration

opaque material part of the transmission mask used in photo-lithography which is not transparent to UV light used for the photoresist exposure; different opaque materials are used in photolithography and X-ray lithography.
transmission mask, photomask, X-ray mask, chrome mask

OPC Optical Proximity Correction; technique used to increase resolution of photolithography; element of computational photolithography.
photolithography, enhancement techniques

open-circuit voltage the voltage of illuminated solar cell measured between its open output terminals.
short-circuit current, solar cell

OPL Optically Pumped Laser.
LASER

optical cavity an integral part of semiconductor laser used to confine light and create standing waves.
semiconductor laser

optical emission spectroscopy, OES a method of endpoint detection in dry etching processes; endpoint is detected based on the changes in the spectrum of radiation emitted by plasma when chemical composition of etched materials changes, i.e. when one material is completely removed.
endpoint detection

optical interconnects planar waveguides formed on the surface of the chip; photons in the waveguide move more efficiently than electrons in metal lines; the use of optical interconnects instead of metal inter-connects in IC chip requires on-chip conversion of electrical signal into optical signal and then back into electrical signal.
interconnect

optical lithography see *photolithography*.

organic contaminants in semiconductor processing mainly hydrocarbons from the resist/etching processes, ambient air, storage environment, shipping boxes, etc.; if not controlled may cause reliability problems in MOS devices as well as may have an adverse effect on the characteristics of metal-semiconductor contacts and epitaxial layers; removal accomplished using strongly oxidizing ambient either wet or dry.
cleaning, contaminant

organic LED, OLED Organic Light Emitting Diode; a LED fabricated using organic semiconductors; a mechanism of light emission is different than in conventional LEDs; in OLEDs radiation is emitted as a result of electron-hole interactions in the thin-film organic semiconductor leading to the formation of exciton; de-excitation of excitons (separation of electrons and holes) results in photon emission; can be formed on flexible substrates which allows flexible displays and lighting panels.
exciton, organic semiconductor, emissive displays

organic semiconductors a class of semiconductor materials based entirely on organic components; low-cost semiconductors formed as a thin-film on any substrate (including flexible substrates); maintain their fundamental characteristics even when significantly bent; by nature much more resistive (in the range of 10^{14} Ω-cm, i.e. almost insulators) than inorganic semiconductors; two key types of organic semiconductors. *(i)* small-molecule and *(ii)* long-chain polymeric; allow formation of TFTs and LEDs; expand applications of semiconductors into new areas, particularly in display technology (flexible displays).
semiconductors, inorganic semiconductors, small molecule organic semiconductor, polymer semiconductor, organic LED, organic TFT, TFT

organic solar cells, OSC solar cells formed using organic semiconductors; also known as excitonic solar cells; low efficiency, but can be formed on flexible substrates; a class of relatively low-cost solar cells.
excitonic solar cells, organic semiconductors

organic solvent a liquid used to dissolve organic matter, e.g. in semiconductor processing acetone can be used to dissolve photoresist.

organic TFT, OTFT Organic Thin Film Transistor; a thin-film transistor formed using organic semiconductors.
thin-film transistor

OSC see *organic solar cell.*

OSF oxidation induced stacking fault; crystal defects that can be formed during high temperature thermal oxidation of silicon.
stacking fault

OSQB organosilsesquioxanes; a class of silsesquioxans; a building block of nanocomposits; depending on chemical composition display semiconductor or dielectric (low-*k*) properties.
low-k dielectric

outdiffusion an undesired effect occurring at high temperature; dopant atoms diffuse in the direction of the concentration gradient from the material featuring high doping level to the material featuring low doping level; common in high temperature epitaxial deposition where it prevents sharp change in dopant concentration between epi layer and the substrate.
autodoping, epitaxy, dopant redistribution

overetching etching process lasting longer than the time required to etch entirely material of interest; may result in undesired lateral expansion of the etched pattern; end-point detection is needed to prevent overetching.
Endpoint, endpoint detection

overlay a superposition of the pattern on the mask and the pattern previously created on the surface of the wafer.
alignment

oxidation chemical reaction converting an element into its native oxide; e.g. reaction of Si with oxygen forms on the silicon surface a layer silicon dioxide SiO_2.
anodic oxidation, plasma oxidation, thermal oxidation

oxidation constants A, B, τ; coefficients used in calculations of oxide thickness in thermal oxidation of silicon; Deal-Grove model; defined for a given temperature of oxidation, oxidizing ambient, and surface orientation; take into account mass transfer in the gas-phase, diffusivity of oxidant in the oxide, rate of surface reaction, as well as thickness of the oxide existing on the surface prior to oxidation.
thermal oxidation, Deal-Grove model

oxidation kinetics a rate at which oxidation reaction occurs; most commonly the term refers to the kinetics of thermal oxidation of silicon.
thermal oxidation, Deal-Grove model

oxidation mask a thin layer of material capable of preventing oxidation of material underneath it even at high temperature; in silicon processing a silicon nitride, Si_3N_4, is acting as a very effective oxidation mask during thermal oxidation.
local oxidation of silicon, silicon nitride

oxide product (in most cases a solid) of oxygen reaction with a given element; usually an insulator (exceptions: e.g. ITO, ZnO); in Si technology the term typically refers to SiO_2 which is a native oxide of Si.
silicon dioxide, zinc oxide

oxide breakdown a change in physical properties of an insulating oxide inflicted by the high electric field; as a result of the breakdown an oxide no longer displays insulating properties.
breakdown, hard breakdown, progressive breakdown, soft breakdown

oxide etching a process causing dissolution of an oxide in a liquid solution (wet etching) or its volatilization in the gaseous ambient (dry etching).
BOE, dry etching, wet etching

oxide fixed charge, Q_f a charge in the Si-SiO_2 structure; located in the oxide in the immediate vicinity of Si surface; does not move or exchange charge with Si, hence, "fixed"; associated with incompletely oxidized silicon.
dangling bond

oxide integrity in common usage the term synonymous with oxide resistance to high electric field stress.
oxide breakdown

oxide mobile charge, Q_m electrically charged species which can move in the MOS gate oxide under the influence of electric field; causes severe instabilities of MOSFET characteristics, e.g. fluctuations of the threshold voltage V_T; Na^+ ions are the most common mobile charges in SiO_2.
ionic conduction

oxide trapped charge charge centers in SiO_2 gate oxide which are electrically activated/de-activated by trapping/de-trapping charge carriers injected into the oxide either from the gate or from the substrate; affects characteristics of the MOSFET.
gate oxide

oxygen a gas; component of air; an element which forms oxides with many elements; an important oxidant in semiconductor technology; the most abundant element.
oxidation

oxygen in silicon oxygen finds its way into silicon during the Czochralski (CZ) single-crystal growth process; in moderate concentration (below 10^{17} cm^3) oxygen improves mechanical properties of silicon wafers; excess oxygen may precipitate in silicon forming electrically active centers.
Czochralski crystal growth, gettering intrinsic

oxygen precipitation see *oxygen in silicon.*

ozonated water water with ozone dissolved in it to increase its oxidizing strength; commonly used in semiconductor processing in cleaning/rinsing operations to control organic contaminants on the wafer surface as well as in the deionized water delivery lines.
ozone, deionized water

ozone a gas comprising of three oxygens, O_3, formed during the reaction of atomic oxygen, O, with molecular oxygen, O_2; very strong oxidizing agent; in semiconductor processing used as organic contamination controlling agent in ozonated water as well as in the gas-phase; unsafe; needs to be handle in a protected environment.
ozonated water, organic contaminants, UV/ozone

P

p concentration of holes in semiconductor (cm^{-3}); also depiction of semiconductor in which holes are acting as majority carriers.
p-n junction, majority carriers

p-channel MOSFET see *PMOSFET.*

***p-i-n*, PIN, junction** semiconductor structure (diode) comprising of *p*- and *n*-type materials with an intrinsic material *(i)* in between; addition of an intrinsic layer changes properties of the *p-n* junction toward improved light absorption characteristics; a key element of the photodetector devices.
intrinsic semiconductor, photodetector, p-n junction

***p-n* junction** a junction involving *p*-type and *n*-type semiconductors brought into contact in order to create a potential barrier; once barrier is established its height is controlled by the voltage applied between *p*- and *n*-type regions; displays diode-like rectifying current-voltage characteristics; in homojunctions the *n*- and *p*-type regions are formed in the same semiconductor typically by locally changing its doping from *p* to *n*, or vice versa; in heterojunctions the *n*- and *p*-type regions are made out of different semiconductors featuring *p*-type and *n*-type conductivity.
p-type semiconductor, n-type semiconductor, potential barrier, forward bias, reverse bias

***p-n* junction isolation** an isolation scheme used in bipolar integrated circuits, i.e. based on bipolar transistors; *n*-type "island" in which *n-p-n* transistor is formed is surrounded by *p*-type material; reverse biased *p-n* junction formed provides isolation.
LOCOS, trench isolation

***p-n-p* transistor** a bipolar transistor with *p*-type emitter and collector and *n*-type base; displays inferior characteristics as compared to its *n-p-n* counterpart due to the lower mobility holes rather than higher mobility electrons acting as minority carriers in the base.
n-p-n transistor, bipolar transistor, lateral transistor

***p*-type conductivity** electrical conductivity of semiconductor dominated by the holes acting as charge carriers.
n-type conductivity

***p*-type dopant** a dopant (alien element) introduced into semiconductor material to render it *p*-type (acceptor).
dopant, acceptor, ion implantation, diffusion

***p*-type semiconductor** a semiconductor material in which concentration of holes is much higher than the concentration of electrons ($p \gg n$); holes are majority carriers and dominate conductivity.
intrinsic semiconductor, n-type semiconductor

package a housing of the chip or discrete device; electrically interconnects chip with outside circuitry; also provides physical and environ-
mental protection of the chip; designed to dissipate heat generated by the
chip; packages are available in the wide variety of designs depending on
the function of the chip/device inside.
DIP, PGA, discrete device

packaging process of enclosing semiconductor chip or discrete device
in a package.
package

PAL Programmable Array Logic.

parallel plate reactor a reactor used in plasma processes in which
grounded and powered electrodes are parallel to each other; plasma is
generated between them and the processed wafers are located or either
grounded or powered electrode.
plasma etching, reactive ion etching

partial planarization surface planarization process which accom-
plishes planar surface only over some parts of the chip surface; as
opposed to global planarization.
planarization, global planarization

Partially-Depleted SOI, PDSOI implementation of the CMOS device
using SOI substrate in such way that the depletion/inversion layer under
the gate is thinner than the Si active layer.
active Si layer, Fully-Depleted SOI, SOI

particle a very small piece of alien material (dust, ultra-small chunks
of silicon or silica, skin flakes, colonies of bacteria, etc.) present in the
process environment; even ultra-small particles (< 0.05 μm) on the Si
surface may cause catastrophic damage to the very high-density chip,
hence, particles must be removed from the wafer surface in the course of
IC manufacturing.
contaminant, particle removal, cleanroom, megasonic cleaning

particle counter an instrument used to detect and measure, depending
on the purpose either *(i)* a number of particles in the volume of air or

process gas or *(ii)* a number of particles per unit area of the wafer surface.
LPD

particle removal process designed to remove particles from the wafer surface; most commonly carried out using wet cleaning (APM) with megasonic agitation; particle removal in the gas-phase is less effective.
APM, cryogenic aerosol, megasonic agitation

pascal pascal (Pa) is the SI unit of pressure; 1 atm = 101325 Pa = 760 torr.
torr

passivation a process rendering semiconductor surface chemically and electrically inactive; term typically refers to the process in the course of which broken bonds at the semiconductor surface are saturated, and hence, de-activated by reaction with selected element; e.g. hydrogen passivates broken Si bonds at the surface; oxide grown on the Si also passivates the surface; sulfur passivates the surface of selected III-V compounds.
hydrogen termination, oxidation

passive element (component) a two-terminal device which displays symmetric, i.e. independent of the direction of the applied bias, characteristics; is not introducing net energy into the circuit; resistors and capacitors are passive elements.
active element

pattern definition a set of processes used to define desired pattern on the wafer surface; in the top-down process it involves film deposition, lithography and etching; the bottom-up process typically involves lithography, surface functionalization and deposition of the thin-film in which desired pattern is reproduced.
etch, lithography, top-down process, bottom-up process

pattern generation, generator transfer of the design layout from the computer files onto the mask by means of direct-write electron-beam lithography; a mask making process.
electron beam lithography, mask, EBIC

158

Pauli exclusion principle no two electrons can occupy the same energy state in the atom at the same time.

PCM see *Phase-Change Memory*.

PCRAM Phase-Change Random Access Memory.
RAM

PDIP Plastic Dual In-Line Package.
DIP

PDP Plasma Display Panel.

PDSOI see *Partially Depleted SOI*

PEALD see *Plasma Enhanced Atomic Layer Deposition*.

PECVD see *Plasma Enhanced Chemical Vapor Deposition*.

PEEM Photo Electron Emission Microscopy.

pentacene small-molecule organic compound displaying semiconductor properties; common organic semiconductor, features electron mobility higher than polymer semiconductors.
organic semiconductors, small molecule- polymer semiconductors

percolation broad term describing slow motion of "something" through the solid; in semiconductor technology refers to the transport of electric charge through insulators, porous or defective in particular; often used to describe progression of the oxide breakdown.
oxide breakdown

permittivity of vacuum, ε_0   8.85 x 10^{-14} F/cm

perovskites materials featuring the same crystal structure as calcium titanium oxide $CaTiO_3$ (perovskite structure); represent a broad class of materials displaying unique characteristics allowing divers applications including manufacture of low-cost, efficient solar cells.
solar cell

PGA see *Pin Grid Array*.

159

Phase-Change Memory, PCM an alternative to conventional electric charge storage based memories; a memory action is based on the change of the phase in, most commonly, chalcogenide glass.
memory cell, chalcogenides

phase-shift mask, PSM a photolithographic mask to which a layer of material featuring desired refractive index and thickness is locally added, or quartz blank is locally thinned in order to shift phase of the UV light passing through the transparent portion of the mask (blank); phase shifting increases resolution of pattern transfer by destructive interference which prevents photoresist exposure in the regions in which it should not be exposed.
enhancement techniques, mask, photolithographic mask

phase shifter a thin layer of material added to the transparent parts of the mask to change the phase of light passing through the mask; with properly selected thickness of the "shifter" the phase can be shifted by 180° and the desired destructive interference accomplished; common phase shifter, e.g. Si_3N_4; alternatively phase shifting is accomplished by locally thinning a quartz blank the UV light is passing through.
phase-shift mask, enhancement techniques

PHEMT, pHEMT Pseudomorphic High Electron Mobility Transistor.
pseudomorphic material, High Electron Mobility Transistor

phonon a "packet", or quantum, of energy of lattice vibrational wave, i.e. periodic variation in the lattice spacing moving across the crystal with a speed of sound; e.g. generated during absorption-emission processes in indirect bandgap semiconductors.
indirect bandgap semiconductor, lattice

phonon scattering phonon's interactions with the lattice when moving across the crystal; adversely alters phonon's transport characteristics.
phonon, scattering, electron scattering

phosphides III-V semiconductor compounds of phosphorus (P, group V) with elements from the group III, e.g. GaP, InP.
antimonides, arsenides, nitrides

phosphine, PH_3 a gaseous compound (toxic) commonly used in silicon manufacturing as a source of phosphorus, P (n-type dopant in Si).
diffusion source, dopant

Phospho-Silicate glass, PSG silica (SiO_2) with phosphorous added to reduce its flow temperature; also used to getter mobile ions in the oxide.
BPSG, gettering

phosphor a luminescent material coating inside walls of the LED light bulb's envelope; absorbs LED's blue light and re-emits longer wavelengths suitable for lighting.
luminescence, LED light bulb, phosphorence, white LED

phosphorence a mechanism producing light by the solids called phosphors as a result of exposure to energy from outside source (e.g. UV light or electron beam).
phosphor

phosphorene also known as black phosphorus (bP), two-dimensional (2D) material displaying distinct, advantageous semiconductor characteristics; energy gap (bandgap) $E_g \sim 0.3$ eV, direct, very high electron mobility; possible in transistor switching applications in which graphene, because of the lack of the energy gap, is of limited use.
black phosphorus, two-dimensional material, graphene, silicene

phosphorus, P group V element; common n-type dopant (donor) in silicon; about 10 times faster diffusant in Si than As and Sb which are other n-type dopants in Si; forms phosphides with group III elements, e.g. GaP, InP.
donor, arsenic, antimony, phosphides

phosphorus tetrachloride, POCl$_3$ a liquid compound commonly used in silicon manufacturing as a source of phosphorus, P (n-type dopant in Si); liquid is vaporized and transferred to the furnace diffusion tube; commonly referred to as "POCL".
diffusion source, dopant, phosphine

Photo-CVD a CVD process in which generation of active reactants is accomplished by photolysis, i.e. by decomposition of gaseous compounds by high-energy radiation, typically UV.
CVD, PECVD

photoconductivity electrical conductivity of semiconductor promoted by its illumination with light featuring energy higher than semiconductor energy gap ($h\nu > E_g$).
photocurrent

photocurrent   a current in semiconductor resulting from its illumination with light featuring energy higher than semiconductor's energy gap.
energy gap

photodetector   semiconductor device the electrical characteristics of which change upon illumination with light featuring energy higher than semiconductor's energy gap; semiconductor (energy gap) is selected to match wavelength of light to be detected.
generation, bandgap, photoresistor, photodiode, phototransistor

photodiode   a diode acting as a photodetector; a two-terminal device with built-in potential barrier; typically a *p-n* junction often in *p-i-n* version; its reverse current changes upon illumination with light featuring energy higher than semiconductor's energy gap.
p-i-n junction, photodetector

photoelectric effect   emission of electrons from the solid as a result of its exposure to high-energy electromagnetic radiation (e.g. ultraviolet or X-ray); foundation of surface analysis techniques such as XPS.
X-Ray Photoelectron Spectroscopy, ESCA

photoelectron   electron emitted from the solid as a result of photoelectric effect; ejected photoelectrons represent measurable electric current.
photoelectric effect

photoemulsion   a material acting as an opaque (non-transparent to UV light) part of the masks used in photolithography; inferior to chrome, used for the same purpose due to the lower optical density (thicker film needed to block off UV radiation), grainy structure (not as sharp edges) and softness making it prone to mechanical damage.
photoemulsion mask

photoemulsion mask   a mask used in photolithography; uses thin film of photoemulsion as an opaque material.
photoemulsion, chrome mask, photomask

photolithographic mask   a mask used in photolithography; see *mask, photolithography*.

photolithography (optical lithography)   a lithography technique which uses UV light to expose resist (photoresist in this case) by through-the-

mask illumination; exposure techniques: either contact or proximity or projection printing; the most common lithography technique in semiconductor manufacturing; using UV wavelength from about 450 nm to 193 nm; projection printing (steppers), phase shift masks, adequate exposure ambient and improved resist technology increase resolution of the photolithographic process; photolithography is capable of resolution down to 10 nm and possibly below with additional modifications.
enhancement techniques, photomask, photoresist, contact-, proximity-, projection printing.

photolithography, computational see *computational lithography.*

photolithography, immersion see *immersion photolithography.*

photoluminescence emission of light from the material stimulated by the absorption of photons (electromagnetic radiation); one of several forms of luminescence.
luminescence, photon, electroluminescence

photomask see *mask, photolithography.*

photon a "packet", or quantum, of electromagnetic energy; flux of photons is constituting electromagnetic radiation; interactions of photon with semiconductor are determined by photon's energy (hv) and bandgap of semiconductor (E_g); energy of incident radiation is absorbed by semiconductor when $hv > E_g$; energy of photons emitted by the direct bandgap semiconductor device as a result of direct band-to-band recombination is equal to the energy gap, i.e. $hv = E_g$.
absorption, recombination, light emitting diode, photonic device

photonic device semiconductor device which operation is based on the interactions of photons in the solid; photon is acting as the information/ energy carrier.
electronic device, semiconductor device, LED, solar cell

photonics technical domain concerned with devices based on photonic interactions.
photon

photoresist photo-sensitive material used in photolithography to transfer pattern from the mask onto the wafer; a viscous liquid deposited on the

surface of the wafer as a thin-film then solidified by low temperature curing; in the areas in which photoresist is exposed to UV light photochemical reactions change its properties (specifically, solubility in the developer); chemically resistant – acts as a mask during subsequent to photolithography etching operations; two types of photoresist: positive and negative.
photolithography, positive resist, negative resist

photoresistor a homogenous semiconductor with two ohmic contacts and acting as photodetector; its electrical conductivity changes upon illumination with light featuring energy higher than semiconductor's energy gap; passive element.
conductivity, passive element, ohmic contact

photostabilization of resist process performed after developing and hard-baking of the resist and before etching or implantation; involves UV irradiation combined with low-temperature ($< 200\ °C$) curing; prevents erosion of resist and outgasing from the resist during subsequent processes.
photoresist, hard bake

phototransistor a transistor acting as a photodetector; based either on heterojunction in which collector-base junction is a *p-i-n* junction acting as photodetector or thin-film transistor with channel exposed to light.
HBT, p-i-n, thin-film transistor

photovoltaic effect generation of the potential difference (open circuit voltage) at the terminals of an unbiased semiconductor diode (almost uniquely a *p-n* junction) as a result of illumination with light featuring energy higher than the energy gap (E_g) of semiconductor; foundation of the semiconductor solar cell operation.
solar cell

photovoltaics term refers to the technical/industrial domain concerned with devices converting energy of sunlight into electrical signal via a photovoltaic effect.
photovoltaic effect

physical etching a process of etching through physical interactions (momentum transfer) between accelerated chemically inert ions (e.g. Ar^+)

and etched solid; sputter etching or ion milling; anisotropic, non-selective; damages the surface.
chemical etching, sputtering, ion milling

Physical Liquid Deposition, PLD   class of thin-film deposition methods using liquid precursors rather than gaseous (CVD) or solid sources (PVD); liquid is applied to the substrate's surface then solidified by thermal curing causing evaporation of the solvent; several PLD methods, including spin deposition and mist deposition, are available.
spin deposition, mist deposition, micro-spray, inkjet printing, CVD, PVD

Physical Vapor Deposition, PVD   a thin-film deposition process carried out in the gas-phase at the reduced pressure; source material is physically transferred in the vacuum to the substrate without any chemical reactions involved; includes evaporation (thermal and e-beam) and sputtering; commonly used to deposit metals.
evaporation, sputtering, Chemical Vapor Deposition, Physical Liquid Deposition

physisorption   the weakest form of adsorption resulting from purely physical attraction (van der Waals force) between the adsorbate and the surface; no chemical bond between the adsorbate and the surface is formed.
chemisorption, van der Waals force

pi **gate**   a 3D MOS gate shape of which in cross-section resembles Greek letter "pi", Π; variation of FinFET technology.
FinFET, gate allaround, omega gate

PIC   Photonic Integrated Circuit; semiconductor based (e.g. InP) mono-lithic integrated circuit for fiber optic networks.
indium phosphide

piezoelectric   a material displaying piezoresistive characteristics; key materials in advanced sensors, actuators, etc.
piezoresistivity

piezoresistivity   change of semiconductor resistivity as a result of the applied mechanical stress; observed also in common semiconductors such as silicon in all crystallographic form, germanium, silicon carbide and others.
sensor

Pin Grid Array, PGA a type of the IC package commonly used to encapsulate very high-density IC chips such as microprocessors; hundreds of leads (pins) come out of the bottom of the package; designed to minimize the distance signal must travel from the chip to designated pin.
MPGA, package

pinch-off a condition occurring in the MOSFET at the source-drain bias at which an inversion layer is "pinched-off" at the drain end of the channel; the pinched-off channel is depleted of the free charge carriers and the transistor current saturates.
MOSFET, saturation region

pinch-off current saturated transistor current (drain current in the MOSFET) flowing in the pinched-off channel; the current in the saturation region on transistor's output characteristics.
pinch-off, Field effect Transistor

pinhole a defect in thin films; a hole in the material with diameter as small as few nm; whether it is in the layer of oxide, organic semiconductor, or resist, a pinhole always has a potential for causing reliability problems.
defect

pionics an area of electronics based on the transport properties of the *"pi"* electron system in graphene; in an extremely confined geometry of graphene the *p* orbitals from neighboring atoms overlap to produce the *pi* bands.
graphene

Piranha clean see *SPM*.

pit in semiconductor terminology the term refers to the physical defects at the semiconductor surface which is basically a hole or an indentation in the material structure.
Crystal Originated Pit, pitting

pitch the center-to-center distance between features of an integrated circuits such as interconnect lines.
integrated circuit

pitting a process of formation of pits at the crystalline semiconductor surface as a result, for instance, of the aggressive chemical treatments.
pit

PL see *photoluminescence.*

planar CMOS a standard CMOS cell featuring horizontal channels induced in the near-surface region of the wafer; as opposed to 3D non-planar CMOS cells in which channels are positioned vertically on the surface of the substrate such as, for instance, in FinFETs.
FinFET

planar defect crystallographic defect also known as area defect; basically an array of dislocations, e.g. grain boundaries, stacking faults.
crystal defects, dislocation

planar *p-n* junction a *p-n* junction formed using a planar process.
planar process

planar process prevailing process mode in semiconductor device manu-facturing; involves processes in which geometry of the doped regions is defined by the pattern created in the masking oxide.
top-down processing

planar transistor a transistor, or other semiconductor device, manu-factured *(i)* using planar process or *(ii)* in which a channel along its entire length is parallel to the wafer surface.
planar process, mesa transistor, planar CMOS, FinFET

planarization a process of rendering the surface of the wafer planar (flat); accomplished either by CMP or by deposition and re-flowing of SiO_2 doped with P and B (BPSG), or by deposition of thick layer of photoresist.
BPSG, Chemical-Mechanical Planarization, global-, partial planari-zation

planarizing resist a thick layer of photoresist deposited to planarize the surface; imaging resist is deposited on top of it.
imaging resist, planarization, partial planarization

Planck constant, *h* 6.63 x 10^{-34} J-s.

Planck-Einstein relation formula states that the energy of a photon E is proportional to its frequency v with Planck constant h as a proportionality constant: $E = h\,v$.
photon

plane channeling in the case of implanted ions a channeling along the crystallographic plane in the single-crystal substrate.
channeling, crystallographic plane, ion implantation

plasma an important medium in semiconductor processing; a self-contained part of the electrical discharge in gases which features equal concentration of ions and electrons; plasma contains electrically active species, but as a whole is electrically neutral; chemically active ions can be extracted from plasma to play variety of functions in semiconductor processing, e.g. in plasma enhanced deposition or plasma etching; chemically inert ions can be used in sputter deposition or ion milling; also a source of ions used in ion implantation.
high-density plasma, PECVD, plasma etching, sputtering, ion milling

plasma afterglow see *afterglow plasma.*

plasma anodization oxidation of the surface of the conductive solid in oxygen plasma.
anodization

plasma ashing removal (volatilization) of photoresist in strongly oxidizing oxygen plasma.
photoresist, plasma

plasma doping see *Plasma Immersion Ion Implantation.*

Plasma-Enhanced Atomic Layer Deposition, PEALD an ALD process in which reactivity of species involved in the deposition sequence is enhanced by plasma.
Atomic Layer Deposition

Plasma-Enhanced Chemical Vapor Deposition, PECVD a process of chemical vapor deposition in which species to be deposited are generated in plasma; as a result, deposition using the same source gases is taking place at lower temperature then in conventional CVD which requires high temperature to break bonds and to release desired species from

input gas molecules; somewhat lower film quality than in the case of pure thermal Low-Pressure Chemical Vapor Deposition (LPCVD).
plasma, CVD, LPCVD

plasma enhancement generation of plasma in the chemical vapor deposition gases for the purpose of lowering deposition temperature and/or generating species featuring enhanced reactivity.
Plasma-Enhanced CVD

plasma etching dry (gas-phase) etching mode in which semiconductor wafer is immersed in plasma containing etching species; chemical etching reaction is taking place at the same rate in any direction, i.e. etching is isotropic; can be very selective; used in those applications in which directionality (anisotropy) of etching in not required, e.g. in resist stripping.
Reactive Ion Etching

Plasma Immersion Ion Implantation, PIII very shallow implantation is taking place when the wafer is "immersed" in plasma containing dopant ions; ions are not accelerated toward the substrate, and hence, penetration depth is very shallow; can be used to form ultra-shallow junctions required by CMOS scaling rules.
ion implantation, scaling rules

plasma oxidation an oxidation process carried out using plasma generated oxygen ions as oxidizing species.
oxidation, thermal oxidation, anodic oxidation

plasmon an optically generated wave of free electrons that can propagate along the surface/interface supporting frequencies in 100 THz range; plasmons are considered for use in future generation chip-level interconnect networks (plasmonic waveguides).

plasmonics technical domain concerned with devices based on plasmonic interactions.
plasmon

plastic ICs integrated circuits fabricated using thin-film organic semi-conductors.
organic semiconductor

platinum silicide, PtSi material for ohmic contacts material in Si technology; resistivity $25 - 35$ $\mu\Omega$-cm; formed at the sintering temperature of $600 - 700$ °C.
silicide

PLCC Plastic Leadless Chip Carrier; IC package.

PLD Pulsed Layer Deposition; in contrast to ALD (Atomic Layer Deposition) in PLD the flow of precursors into the deposition chamber is continuous while energy needed to stimulate deposition reaction is pulsed.
Atomic Layer Deposition

PLED Plastic Light Emitting Diode; see *OLED*.
organic semiconductor

plug vertical connection between metal lines in multilevel interconnect scheme; commonly tungsten (W) is used for this purpose.
tungsten, via

PMMA Poly(Methyl MethAcrylate); component (matrix) of advanced resists; can act on its own as a positive deep UV resist; also used as an electron-beam resist.
photoresist

PMOLED Passive Matrix display constructed using organic light emitting diodes.
AMOLED, OLED

PMOSFET *p*-Channel MOSFET; a Metal-Oxide-Semiconductor Field Effect Transistor with *p*-type channel, i.e. device in which holes are responsible for conduction in the channel; built into *n*-type substrate with p^{++} doped source and drain; in spite of inferior to NMOSFET performance needed in CMOS technology where it forms with an NMOSFET a complimentary pair.
channel, NMOSFET, CMOS

POA Post-Oxidation Anneal; thermal treatment applied in order to improve properties of an oxide and its interface with silicon.
Rapid Thermal Annealing

POCL   see *phosphorus tetrachloride POCl₃*.

point defect   highly localized imperfection of a crystalline structure; affects the periodicity of the crystal mostly in, or around, one unit cell; point defect can be a vacancy, or interstitial, or substitutional defect.
crystal defects, interstitial defect, substitutional defect, vacancy

Point of Use, POU   term typically refers to the process of generation of chemicals for wet processing at the point of use, i.e. in the tool where they are used; e.g. injection of ozone into deionized water to fabricate ozonated water.
ozonated water

Poisson equation   fundamental relationship defining potential and electric field distribution in semiconductors.

polar semiconductor   doped (extrinsic) semiconductor featuring distinct conductivity type: *n*, or *p*.
n-type semiconductor, p-type semiconductor, ambipolar semiconductor

POLED   Polymer Organic Light Emitting Diode (OLED) fabricated using polymer based organic semiconductor; feature somewhat lower quality, but lower cost than SMOLEDs.
polymer semiconductor, small-molecule semiconductor, SMOLED

polishing   an operation applied to either reduce roughness of the wafer surface or to remove excess material from the surface; typically a mechanical process using chemically reactive slurry.
CMP, CP-4 etch

poly depletion   term refers to an undesired effect in poly-Si MOS gate contacts which results from oxidation of poly-Si gate material with oxygen originating from the gate dielectric; a reason for elimination of poly-Si gate contacts in MOSFETs featuring high-*k* gate dielectrics.
gate dielectric, HKMG, poly-Si gate

poly-Si gate   polycrystalline Si gate commonly used as a gate contact in MOSFETs; used instead of metal gates because work function of poly-Si matches work function of Si substrate much closer than metals, hence, threshold voltage of transistor is significantly lower; not used in conjunction with high-*k* dielectric MOS gates.
gate, gate contact, polycrystalline silicon, poly depletion

polycide gate   poly-Si gate with its top region converted into a silicide.
poly-Si gate, silicide

polycrystalline material   a crystal in which   long-range order exists only within limited in volume grains; grains are randomly connected to form a solid; size of the grains varies depending on material and method of its formation.
amorphous material, single-crystal, multicrystalline material, grain

polycrystalline silicon, poly-Si   crystalline Si in which long range order is maintained only within limited grains; heavily doped poly-Si is commonly used as a gate contact in silicon MOS and CMOS devices; as a cheaper than single-crystal form of silicon can also be used to fabricate thin-film transistors and solar cells.
amorphous Si, poly-Si gate

polyimide   a type   of an organic   insulator   displaying   dielectric constant $k \sim 2$.
low-k dielectrics

polymer semiconductor   a polymer based organic semiconductor composed of the long chains of organic molecules.
small-molecule semiconductor, organic semiconductor

Poole-Frenkel conduction   see *Poole-Frenkel emission.*

Poole-Frenkel emission   electrical conduction mechanism in insulators; involves defect/impurity related electron traps in the material; trapped electrons can escape by thermal emission; current flows due to electrons "jumping" from trap to trap in the presence of an electric field.
hopping

POR   Process of Record.

porous dielectric   porosity (air gaps) is created in an inter-layer dielectric in the multi-level metallization scheme to reduce its dielectric constant k.
ILD, low-k dielectric, nanoglass, multilevel interconnects

porous silicon   silicon with pores (air gaps, voids) in its structure; at the high void fraction remaining silicon displays quantum confinement effects leading to the widening of the energy gap and resulting visible

red-orange light emission; due to its low electroluminescence efficiency this effect is not exploited in commercial devices; porous Si is explored in the range of other applications.
electroluminescence

positive charge dielectric a thin-film dielectric material used in semiconductor technology which features defects that are inherently predominantly positively charged; e.g. thermal SiO_2.
negative charge dielectric, thermal SiO₂

positive resist a resist (e.g. photoresist) which is initially insoluble in the developer and becomes soluble as a result of irradiation; chain scission takes place in the resist structure during irradiation; allows higher resolution than negative resist, but is less sensitive, i.e. requires longer exposure time which hampers the process throughput.
developer, negative resist, photoresist, chain scission

post-CMP cleaning particularly demanding cleaning routine needed to removed products of CMP process; wet cleaning involving mechanical interactions such as brush scrubbing.
brush scrubbing, CMP, slurry, wet cleaning

post-implantation anneal an anneal applied after implantation *(i)* to activate implanted dopants and *(ii)* to repair implantation damage; must be carried out at the low thermal budget (e.g. high temperature for a very short time) to avoid dopant redistribution.
ion implantation damage, dopant activation, dopant redistribution, RTA, thermal budget

potential barrier an increased potential at the junction between two materials featuring different work function; e.g. *p*-type and *n*-type semiconductor (*p-n* junction) or metal-semiconductor contact (Schottky diode); height of potential barrier changes with bias voltage (reverse bias - high potential barrier, forward bias - low potential barrier) which makes junction to display diode-like current-voltage characteristics.
metal-semiconductor contact, p-n junction, diode

POU see *Point of Use.*

power-delay product a measure of semiconductor device performance in digital applications; lower power-delay product means that power is

better "translated" into speed of operation; a figure of merit in digital ICs which shows that power needed to switch (lower the better) is at least as important to digital performance as switching speed.
digital device, digital IC

power device   semiconductor device (diode, transistor) designed to operate at high voltage and high current.
DMOS, IGBT, U-MOSFET, thyristor

ppb   parts per billion.

ppm   parts per million.

PRE   Particle Removal Efficiency; expressed in %; related to a particle size range; needs to be close to 100% for the particle removal method to be commercial manufacturing ready.
particle, particle removal

pre-damage implants   implantation of silicon or germanium into silicon surface region prior to source/drain doping implants; the goal is to pre-amorphize Si and by doing so to suppress channeling during subsequent source/drain doping implants; better control of the depth of implanted junctions is achieved.
channeling, shallow junction, amorphous silicon

pre-deposition   the first step in semiconductor doping by diffusion; process of thermal oxidation of silicon in the ambient containing dopant atoms; heavily doped oxide formed as a result is acting as a source of dopants; same as constant-source (unlimited source) diffusion.
diffusion, drive in, limited source diffusion

preferential etching   etching occurring preferential along selected crystallographic plane(s) in the crystalline solid; typically carried out in the liquid-phase (wet etching).
etching, wet etching, crystal plane

pressure units   Atmosphere (Atm) = standard sea level pressure = 760 torr; Torr (torr) = 1 mm of mercury; Pascal (Pa) = 0.0075 torr; Pounds per Square Inch (psi) = about 51.7 torr.

PREVEIL   Projection Exposure with Variable Axis Immersion Lenses; variation of the e-beam projection lithography.
Electron Projection Lithography

process diagnostics characterization of material or device for the purpose of identification of the nature of malfunctions of the manufacturing process; methods of chemical/physical/electrical material characterization are used.

process integration term usually refers to the integration of two or more individual processing steps in semiconductor manufacturing into a single self-contained tool.
cluster tool

process monitoring determination of the selected characteristics of the surface/material at various stages of the device manufacturing sequence; carried out for the purpose of evaluation of the process performance; goal: as early as possible detection of process malfunction; the most effective when carried out in-line, in real-time.

Programmable Read-Only Memory, PROM nonvolatile memory; basically a Read Only Memory equipped with fuses which can be blown in order to write-in specific information.
nonvolatile memory

progressive breakdown gradual degradation, and eventual breakdown, of the very thin gate oxide in MOS devices under the prolonged electric field stress.
oxide breakdown, gate oxide, TDDB

projected range term used in ion implantation terminology; projected range (R_p) is a distance from the surface of implanted material to the plane where implanted ions reach maximum concentration (concentration peak); represents depth of implantation; controlled by implantation energy.
ion implantation

projection printing a photoresist exposure technique used in photolithography in which image of the mask (reticle) is projected on the surface of the wafer through the complex system of image correcting lenses; assures the highest resolution among all exposure methods;

implemented using steppers in which resist on the surface of the wafer is exposed step-by-step (step-and-repeat process) as opposed to full-field exposure in which the entire wafer is exposed in one shot.
contact printing, proximity printing, stepper, reticle

PROM see *Programmable Read-Only Memory.*

protocrystalline material term refers to the distinct crystal phase occurring during the growth of hydrogenated Si (Si:H); during the film growth this phase evolves into a microcrystalline Si:H; features unique electronic and optical properties.
amorphous silicon, hydrogenated silicon

proximity effect deleterious effect in e-beam lithography; scattering of incoming electrons and emission of secondary electrons in irradiated resist; proximity effect is responsible for the size of the exposed resist area being larger than the diameter of the incident electron beam; limits resolution of e-beam lithography.
electron-beam lithography

proximity printing a photoresist exposure technique used in photo-lithography in which during exposure mask is very close to (distance in the range of 20 μm), but not in physical contact with a resist on the wafer surface; mask damage is less likely than during contact printing, but resolution of pattern transfer is limited to about 2 μm.
contact printing, projection printing

pseudomorphic material term refers to the lattice-mismatched hetero-structures; pseudomorphic film is a layer of single-crystal material deposited on single-crystal substrate featuring slightly different chemical composition, and hence, slightly different lattice constant; lattice mismatch is accommodated by strain in the film; stays pseudomorphic only when thinner than certain critical thickness (h_c) above which stress in the film is released by the formation of dislocation.
lattice mismatch, strained layer superlattice

PSG see *Phospho-Silicate glass.*

PSM see *Phase-Shift Mask.*

puller a high precision mechanical apparatus designed to pull (with rotation) single-crystal from the melt during the CZ crystal growth process.
crystal pulling, Czochralski crystal growth

Pulsed Laser Deposition, PLD a thin-film deposition technique of potential use in the range of semiconductor processing related applications.

pumping speed expressed in liters/second (L/s) or cubic feet/minute (cfm); a parameter defining performance of the vacuum pump used in semiconductor process tools.
roughing pump, turbomolecular pump

PV Photovoltaics; *see photovoltaics.*

PVD see *Physical Vapor Deposition.*

PZT Lead-Zirconate-Titanate; ferroelectric material featuring dielectric constant $k > 1000$; between others used to increase capacitance of storage capacitors.

Q

Q_{bd} see *charge-to-breakdown.*

QD LED Quantum Dot Light Emitting Diode; efficient, bright and color saturated LED formed using semiconductor quantum dots; takes advantage of high photoluminescence efficiency featured by compound semiconductor quantum dots.
Light Emitting Diode, nanocrystalline quantum dot, photoluminescence

QFP Quad Flat Package; variation of the quad package which along with dual in-line package (DIP) is among the most common types of packages used in IC technology.
quad package, dual-in-line package

quad package type of an IC package with leads located on four sides of the package; allows packaging of more complex devices than the DIP (Dual In-Line Package) with leads on the two sides only.
dual in-line package

quantum device semiconductor device which operation is based on the quantum effects resulting from the extreme geometrical confinement of device features and which operation cannot be described by the laws of classical physics.
quantum well

quantum dot (nanodot) see *nanocrystalline quantum dot.*

quantum efficiency see *external quantum efficiency.*

quantum well, QW an ultra-thin layer of narrower bandgap semiconductor sandwiched between two layers of wider bandgap semiconductor; creates a potential well within the materials system; realized using III-V semiconductors, e.g. GaAs and $Al_xGa_{1-x}As$; within the quantum well electrons are 2-dimensionally confined and display characteristics distinctly different than in the 3-dimensional crystal; used in high performance transistors and laser diodes.
High Electron Mobility Transistor, quantum well MOSFET, quantum well laser, bandgap engineering

quantum well laser a laser diode in which an ultra-thin undoped region (quantum well) acting as active region is formed between *n*- and *p*-type regions; due to the quantum confinement of the active region the wavelength of emitted light is controlled by the thickness of the well rather than by the width of the energy gap of material from which it is composed; highly efficient short wavelength emission results; common materials configuration: AlGaAs-GaAs-AlGas.
semiconductor laser, quantum well

quartz a single-crystal form of SiO_2; in semiconductor technology used as a blank (transparent part) in photomasks; due to the high melting point (1670 °C) and purity, quartz is an important element in semiconductor thermal processes infrastructure (furnace tubes, boats, cassettes, etc.)
photomask, blank, furnace

quasi-Fermi level the concept of quasi-Fermi level (quasi-Fermi energy) has no direct thermodynamic interpretation; different for electrons and holes; this concept is used in the description of the excess carrier distribution when external potential is applied to semiconductor

and Fermi potential can not be used to describe distributions of both electrons and holes.
Fermi level, Fermi potential

quasi-neutrality a condition in which in thermal equilibrium concentration of free electrons in n-type semiconductor is not exactly equal to the concentration of donor dopants, i.e. $n_0 \sim N_d$; under this condition an electric field exists in semiconductor.
thermal equilibrium

quasi-static C-V capacitance-voltage measurement of MOS capacitor performed at the very low frequency of a.c. signal, e.g. 50 Hz; combined with high-frequency (e.g. 1 MHz) C-V measurements provides quantitative information regarding electronic properties of the dielectric-semiconductor interface in MOS gate stack.
capacitance-voltage measurement

quaternary semiconductor semiconductor compound comprised of four elements; most commonly involves elements forming III-V semiconductors, e.g. InAlGaAs.
binary semiconductor, ternary semiconductor, AIII-BV semiconductors

QW see *quantum well.*

R
R2R see *roll-to-roll process.*

R-metal MOS gate see *R-MOSFET.*

R-MOSFET a MOSFET in which gate contact is made out of refractory metal (R-metal) such as tungsten (W), molybdenum (Mo), or tantalum (Ta); advantage: R-metals withstand high temperature, i.e., wafer can be subjected to elevated temperature treatments after gate formation; disadvantage: relatively low conductivity of R-metals, incompatibility of work function with silicon, and process complexity.
gate contact, work function difference, refractory metal

radial *p-n* junction   as opposed to a planar *p-n* junction a junction that is implemented in the nanowire (nanorod) in the radial direction; conceived to improve performance of solar cells.
p-n junction, planar p-n junction, nanowire

radiant heating   heating of semiconductor wafer by radiation emitted by a solid at very high temperature; very strong IR component; generated by high-power lamps or high resistivity wire through which high density current is flowing; in the latter case under the steady-state conditions radiation source and the heated wafer are in thermal equilibrium; in the case of the typically short lamp heating cycles no thermal equilibrium is reached; e.g. batch furnaces employed in semiconductor manufacturing use thermal equilibrium radiant heating.
inductive heating, resistance heating, rapid thermal processing

radiation damage   a damage (creation of electrically active defects) inflicted on semiconductor material/device by high energy (very short wavelength) radiation; e.g. X-rays exposure creates active defects in silicon dioxide, SiO_2.
defect, silicon dioxide

radiation hardness   term refers to the resistance of the material system to high energy radiation (e.g. γ-rays, X-rays); in some materials (e.g. in SiO_2) exposure to high energy radiation creates defects responsible for its deteriorated characteristics; an issue of particular concern in the case of outer space and/or military applications.
silicon dioxide

radiation wavelength λ - energy E conversion λ (μm) $= 1.24/E$ (eV).

radiative recombination   electron-hole recombination process in which energy is released in the form of electromagnetic radiation (photons); in the case of band-to-band recombination in direct bandgap semiconductors the energy of radiation released corresponds to the bandgap of semiconductor.
photon, recombination, direct bandgap semiconductor, band-to-band recombination, phonon

radicals   highly reactive dissociated atoms and molecules present in the electrically discharged gas; actively involved in the gas-phase etch reactions.
glow discharge, dry etching

Radio Frequency, RF   a broad range of frequencies from about 3 KHz to 300 GHz; in semiconductor terminology term "RF" typically refers to the frequency of 13.56 MHz used in semiconductor process equipment (plasma generators).
microwave plasma

Radio Frequency, RF, plasma   plasma generated at 13.56 MHz.
microwave plasma

raised source-drain   same as elevated source/drain; feature of the advanced CMOS/MOSFET design; source and drain regions are raised (additional layer of Si is deposited epitaxially) with respect to the plane of the gate oxide-Si interface; implemented to improve performance of the short channel transistors by decreasing series resistance of the very shallow source and drain regions.
drain engineering, elevated drain, elevated source

RAM   Random Access Memory, a memory cell designed to store information (data) temporarily.
DRAM, SRAM

ramp voltage oxide breakdown, E_{bd}   a measure of the reliability of oxides in MOS gates; gate voltage is gradually increased until oxide breaks down (uncontrolled current flows across the oxide) and oxide breakdown voltage is determined; oxide breakdown field is a ratio of breakdown voltage over oxide thickness.
GOI, oxide breakdown, charge-to-breakdown

random motion   a motion of free charge carriers in semiconductor in the absence of an electric field; carriers are moving and subject to collisions, but net transport of the charge in any specific direction (drift) does not occur.

Random Telegraph Noise, RTN   electronic noise occurring in ultra-small semiconductor devices; observed as fluctuations of the device resistance in the form of random switching between typically two discrete values; also known as burst noise.
noise, flicker noise

Rapid Thermal Annealing, RTA an annealing process carried out for a very short time; typically performed for the purpose of improving properties of materials and/or device structures.
Rapid Thermal Processing, thermal budget

Rapid Thermal Chemical Vapor Deposition, RTCVD thermally enhanced CVD process carried out at high temperature, but for a short period of time.
CVD, thermal budget, Rapid Thermal Processing

Rapid Thermal Cleaning, RTC brief increase of the wafer temperature, usually by illumination using high power lamps, in ambient air for the purpose of volatilization of surface organic contaminants and moisture; typically carried out at temperatures below 300 °C in oxidizing ambient, e.g. air at atmospheric pressure.
lamp cleaning, Rapid Thermal Processing

Rapid Thermal Nitridation, RTN high temperature, very short (typically < 60 sec) anneal of oxidized Si wafer in nitrogen containing gas such as NH_3, NO, or N_2O; carried out for the purpose of adding nitrogen to the oxide, i.e. oxide nitridation.
boron penetration, nitrided oxide, Rapid Thermal Processing

Rapid Thermal Oxidation, RTO growth of an oxide on Si surface during high temperature, short time (typically < 60 sec) exposure to oxygen containing ambient.
thermal oxidation, thermal oxide, Rapid Thermal Processing

Rapid Thermal Processing, RTP process that rapidly increases temperature of the wafer and maintains it at the target temperature for a short period of time (typically less than 60 sec, hence, "rapid"); heating is accomplished using high-power lamps (typically halogen-quartz) installed in batches in which power of each lamp is controlled individually; RTP allows high temperature processing at the low thermal budget; used mostly to process single wafers; see *RTA, RTC, RTN,* etc.
furnace, thermal budget, single-wafer processing

Rapid Thermal Silicidation, RTS a brief thermal treatment (sintering) needed for the metal in contact with Si to form an alloy (silicide), e.g. as a result of the RTS process, Co in contact with Si will form $CoSi_2$.
silicidation, silicide, sintering, Rapid Thermal Processing

182

raster scan scanning mode in which beam is moving back and forth over the entire substrate; beam is turned on over designated area and then turned off until it will arrive at the next designated area; one of the scanning modes in e-beam lithography; rarely used in manufacturing because of the long time it takes to "write" a pattern on the wafer surface.
variable shape beam, vector scan

Rayleigh equation defines resolution R (nm) of the photolithographic pattern transfer process; $R = k\lambda/NA$ where k is the process factor, λ is the wavelength of the resist exposing light and NA is a numerical aperture.
numerical aperture

RBS see *Rutherford Backscattering*.

RCA clean using the name of the company where this wafer cleaning methodology was developed the term refers to a specific wet cleaning recipe introduced at RCA in early '70; originally introduced as a two-step process involving RCA-1 and RCA-2 solutions; variations of the basic RCA wafer cleaning sequence are successfully used until today.
wet cleaning

RCA-1 same as APM and SC-1; see *APM* for specific information.

RCA-2 same as HPM and SC-2; see *HPM* for specific information.

reactive evaporation same concept as reactive sputtering, but involving evaporating species rather than sputtered species.
evaporation, reactive sputtering

Reactive Ion Beam Etching, RIBE etching process which uses a beam of chemically reactive ions; combines physical and chemical interactions with the etched material; in contrast to etching using beam of chemically neutral ions such as Ar^+ which involves physical etching only.
ion milling, Reactive Ion Etching

Reactive Ion Etching, RIE variation of plasma etching in which during etching semiconductor wafer is placed on the RF powered electrode; wafer takes on potential which accelerates etching species extracted from plasma toward the etched surface; as a result, chemical etching reaction is preferentially taking place in the direction normal to

the surface, i.e. etching is more anisotropic than in plasma etching, but is less selective; leaves etched surface damaged; the most common etching mode in semiconductor manufacturing.
MERIE, DRIE, plasma etching

Reactive Ion Etching damage   a physical damage to the surface and near-surface region of the material exposed to RIE etch chemistries; results from the energetic interactions between etching species and the surface during reactive ion etching; it also results in the incorporation of species involved in the etch reaction (hydrogen and carbon in particular) into the sub-surface region of the substrate wafer subjected to etching.
damage

reactive sputtering   sputter deposition process in which species sputtered off the target material are chemically reacting with other species in the gas mixture to form a desired compound to be deposited; e.g. sputtering of Si in plasma containing oxygen will result in deposition of SiO_2.
sputtering

Read-Only Memory, ROM   a memory cell with permanently stored information and with no ability to accept new information.
RAM

recessed gate   non-planar configuration of the gate contact implemented in heterostructure compound semiconductor transistors (e.g. HEMT) to improve control over transistor's threshold voltage and transconductance.
High-Electron Mobility Transistor, transconductance

recombination   process resulting in annihilation of free charge carriers in semiconductor; reversal of the generation process; in band-to-band recombination electron recombining with hole looses energy and "drops" from the conduction band to the valence band.
generation, radiative recombination, band-to-band r., Auger r., trap-assisted r.

recombination center, recombination site   an impurity or electrically charged point defect in the bulk or at the surface of semiconductor in which minority carrier is captured and then recombined with subsequently captured majority carrier.
point defect, recombination, trap assisted recombination

recombination current the current resulting from the flow of charge carriers that recombine in the space-charge region of the forward biased *p-n* junction.
generation current, space-charge region, p-n junction

recombination lifetime an average time needed for an electron-hole pair to recombine; represents a decay of excess minority carriers due to recombination.
generation lifetime, excess carriers

recombination rate the rate at which electron-hole pairs are recombining (number/cm^3 sec).
generation rate

rectifying contact in general, a contact displaying diode-like asymmetric current-voltage characteristics, i.e. allowing high current to flow across under the forward bias condition and blocking current off under the reverse bias; this behavior is controlled by the bias voltage dependent changes of the potential barrier height in the contact region; Schottky diode and *p-n* junction diode represent rectifying contacts.
Ohmic contact

redistribution see *dopant redistribution.*

reflection a light ray travelling through the medium and arriving at the surface of the other medium is returned without penetrating it.
refraction

reflective mask see *mask, reflective.*

reflective notching distortion of the pattern created in the layer of photoresist by reflection of the UV light from the wafer surface.
photolithography

refraction change of the direction of propagation of the light ray passing from one medium (refractive index n_1) to another medium featuring different density (refractive index n_2); refracted light ray is transmitted through the solid as opposed to be reflected from its surface.
reflection

refractive index *n* unit-less parameter determining optical properties of solids, liquids, and gases; defined as a ratio of the speed of light in free

space over the speed of light in any given medium; $n \geq 1$, refractive index of air $n = 1$, water $n = 1.34$, SiO_2 $n = 1.46$, Si $n = 3.42$.
dispersive medium

refractory metal a metal featuring very high melting point, e.g. tungsten featuring a melting point of 3422 °C.
R-MOSFET

regioregular polythiophene an organic compound displaying semi-conductor properties.
organic semiconductor

relaxation time a time needed for the charge carriers in semiconductor to return to equilibrium concentration after temporary disturbance (by, for instance, brief illumination) has been terminated.

release process see *MEMS release*.

remote plasma, downstream plasma terms describe plasma process in which wafer is located away from plasma, and hence, is not directly exposed to plasma and plasma generated radiation; desired reactions (e.g. etching) are implemented by extracting ionized species from plasma and directing them toward the wafer; remote plasma process results in less surface damage than standard plasma process as plasma generated ions are energetically relaxed arriving at the surface of the wafer.
direct plasma, plasma

resist material sensitive to irradiation i.e. it changes its chemical properties when irradiated; in the form of thin-film used as a pattern transfer layer in lithographic processes in semiconductor manufacturing and often also as a mask during subsequent etching.
photoresist, e-beam resist, X-ray resist

resist ashing process of resist removal (resist stripping) in strongly oxidizing gaseous atmosphere; typically in oxidizing plasma; see also *resist stripping*.

resist stripping process in which resist is removed from the surface when it is not needed any longer, i.e. after completion of etching operation, or ion implantation; carried out using either wet or dry strongly oxidizing chemistries; can be altered by changes in chemical

composition of the resist occurring during processing, e.g. during ion implantation.
barrel reactor, SOM

resistance heating see *Joule heating.*

resistivity material parameter determining semiconductor's ability to conduct current; resistance per unit area and unit length; expressed in Ω-cm; the reciprocal of conductivity; in contrast to metals and insulators resistivity of semiconductors can be controlled within several orders of magnitude by changing dopant concentration.
conductivity

resolution in semiconductor terminology the term refers to the precision with which lithographic process transfers pattern from the mask to the layer of resist; smaller geometries that can be defined, higher the process resolution; determined by several factors related to the exposure tool (shorter exposure wavelength λ and larger numerical aperture *NA* increase resolution), resist, and masks used.
Rayleigh equation, enhancement techniques, lithography

resonant tunneling an effect occurring primarily in quantum-well double barrier heterostructures; for a given thickness of the well and at the certain bias voltage resonant tunneling causes rapid increase of the current flowing across the structure.
tunneling, quantum well

RET Reticle Enhancement Technique used to increase resolution of photolithoraphy.
resolution, reticle

reticle a name for the mask used in projection lithography where pattern on the surface of the reticle is larger than the pattern created in the layer of photoresist on the wafer surface; in contrast, in the case of contact/proximity printing the mask pattern is transferred to the photoresist without reduction.
mask, projection printing, stepper, full-field exposure, contact printing

retrograde well an approach to the formation of "well" in CMOS structures; highest concentration of dopant implanted in the well is

located at certain distance from the surface; denser, less susceptible to latch-up CMOS device result.
CMOS, well

reverse bias a bias at which heights of the potential barrier at the *p-n* or metal-semiconductor junction is increased and current flow from one region to another is restricted.
forward bias, potential barrier, saturation current

reverse breakdown voltage the value of the revers bias voltage at which permanent breakdown of the *p-n* junction occurs and junction is no longer capable of blocking the current flow.
breakdown voltage

reversible breakdown a non-catastrophic breakdown of the *p-n* junction or the MOS gate dielectric, i.e. a breakdown which can be reversed by reducing the bias voltage.
avalanche diode, soft breakdown

RF see *Radio Frequency*.

RGA Residual Gas Analyzer; an instrument installed in the vacuum process tools to monitor composition of gaseous ambient during semiconductor wafer processing.

RHEED Reflection High Energy Electron Diffraction variation of HEED; in RHEED electrons arrive at the analyzed surface at the grazing angle, and hence, information on the crystallographic structure specific to the surface can be obtained.
HEED, LEED

RIBE see *Reactive Ion Beam Etching*.

RIE see *Reactive Ion Etching*.

rinsing a process in which wafer is immersed in deionized water in order to stop chemical reaction initiated during preceding operation and to remove products of this reaction from the surface.
deionized water

RMOS Resistor Load MOS.

roll-off the term is used in reference to the reduction of the flat-band voltage (V_{FB}) of the PMOSFET with high-k gate dielectric and high-work function metal gate; occurs when an equivalent oxide thickness (EOT) is being reduced.
EOT, PMOSFET, high-k dielectric, flat-band voltage

roll-to-roll process, R2R the process which forms functional electronic and/or photonic devices on the flexible plastic or metal foil continuously moving from roll-to-roll while device features are formed.
flexible substrate

rms Root Mean Square; often used as a measure of variations of surface topography (surface roughness) determined by Atomic Force Microscopy.
AFM, surface roughness

ROM see *Read-Only Memory*.

Roots pump, Roots blower high efficiency roughing pump; used in oil vapor-free vacuum systems.
roughing pump, mechanical pump

ROST Rapid Optical Surface Treatment see *lamp cleaning* and *rapid thermal cleaning*.

rotagoni drying Marangoni drying combined with rotation of the wafer; compatible with single-wafer processing.
Marangoni drying, single-wafer process

roughing pump a vacuum pump designed to reduced pressure inside the semiconductor process tool from atmospheric to militorr range; used before pump operating in the lower pressure regime can be turned on, oil-free (dry pumps) are preferred.
Roots pump, mechanical pump, dry pump

roughness lack of planarity of the solid surface; the high quality semi-conductor surfaces feature atomic level roughness, i.e. less than 0.1 nm peak-to-valley.
AFM, surface roughness

roughness scattering scattering of the charge carriers moving in the vicinity of semiconductor surface caused by surface roughness.
scattering, surface roughness

RTA   see *Rapid Thermal Annealing.*

RTC   see *Rapid Thermal Cleaning.*

RTCVD   see *Rapid Thermal Chemical Vapor Deposition.*

RTN   see Rapid *Thermal Nitridation.*

RTN   see *Rapid Telegraph Noise.*

RTO   see *Rapid Thermal Oxidation.*

RTP   see *Rapid Thermal Processing.*

RTS   see *Rapid Thermal Silicidation.*

Rutherford Backscattering, RBS material characterization method providing information on the composition and depth distribution of elements in the investigated solid; solid is bombarded with high energy helium, He, ions; information regarding composition of the material is obtained by determining energy of the backscattered He ions.

S

sacrificial oxide *(i)* oxide formed on the Si surface in the course of MEMS device manufacturing to provide mechanical support for some of its parts; oxide is then etched away to allow movements of the part (MEMS release process); *(ii)* silicon dioxide thermally grown on the Si surface and then etched away to reveal Si surface less damaged than original one.
MEMS release

SACVD   see *Selective Area Chemical Vapor Deposition.*

safe operating area   operation of the power MOSFET without exceeding maximum rated current, voltage, and power.
power device

"SALICIDE" process   Self-Aligned Silicide process; process in which silicide contacts are formed only in the areas in which metal deposited to

form silicide is in direct contact with silicon, hence, are self-aligned; commonly implemented in MOS/CMOS processes using poly-Si gate in which silicide ohmic contacts to the source, drain, and poly-Si gate are formed and aligned in the course of a "SALICIDE" process.
silicide, ohmic contact

SAM   Scanning Auger Microscopy, see *Auger Electron Spectroscopy.*

SAM   see *Self Assembled Monolayer.*

SAMOS   Self-Aligned MOS.

sapphire   single-crystal aluminum oxide, Al_2O_3; highly chemically resistant, hard material, transparent to UV; can be synthesized and processed into various shapes including large wafers used in semiconductor engineering as substrates for silicon (Silicon-on-Sapphire, SOS) and gallium nitride.
aluminum oxide, Silicon on Sapphire, gallium nitride

saturation current   current in the reverse biased *p-n* junction; independent of bias voltage; depends on concentration of minority carriers and quality of semiconductor.
p-n junction

saturation region   an operating mode of the transistor in which its output current saturates at its highest value for any given bias conditions; transistor is fully-"on".
cut-OFF region

saturation velocity   maximum velocity of charge carriers in semiconductor; electrons and holes velocity in semiconductor increases as a function of electric field up to certain maximum value (typically at the level of $\sim 10^6 - 10^7$ cm/s) which cannot be exceeded due to the excessive carrier scattering at the very high electric field; electric field at which saturation occurs is different for electrons and holes and depends on semiconductor; e.g. for electrons in Si $\sim 10^5$ V/cm, and in Ge $\sim 3 \times 10^3$ V/cm.
drift velocity, scattering

SC-1   "standard clean" 1 see *APM.*
cleaning, RCA-1

SC-2 "standard clean" 2 see *HPM*
cleaning, RCA-2

scaling term refers to the continued downsizing of the key geometrical features of the transistors in logic and memory integrated circuits.
gate scaling, logic IC, memory IC

scaling rules design rules which must be followed while scaling down geometry of transistors and interconnect lines comprising an integrated circuit; arbitrary reduction of key geometries, e.g. channel length in MOSFETs, may results in the dysfunctional devices, e.g. short-channel effects in MOSFET.
short-channel effects

SCALPEL Scattering with Angular Limitations Projection Electron Lithography; variation of projection electron lithography (EPL).
electron-beam lithography

Scanning Electron Microscopy, SEM imaging method with lateral resolution better than 10 nm; focused beam of electrons is scanned across the sample; image constructed based on the detection of secondary electrons current; studied material must be coated with a conductive film; very common, very useful in semiconductor engineering to visualize very small geometrical features on the wafer surface.

Scanning Transmission Electron Microscopy, STEM combines capabilities of Transmission Electron Microscopy (TEM) and Scanning Electron Microscopy (SEM) to offer superior spatial resolution.
Transmission Electron Microscopy

Scanning Tunneling Microscopy, STM method allowing visualization of conductive solid surfaces with atomic resolution; conductive tip is scanned across the surface of the sample; information about the physical features of the surface is based on the measurement of the tunneling current between the tip and the surface atoms.
tunneling

scattering a process responsible for an electron in a certain state (defined by its crystal momentum) suddenly moving into a different state; a result of the electron interactions with host atoms in the lattice, dopant atoms, as well as defects; fundamental effect defining electron

transport in semiconductors by affecting carriers mobility; phonons moving across solids are also subject to scattering.
mobility, phonon scattering, Coulomb scattering, roughness scattering, dopant scattering

sccm standard centimeter cube per minute; common unit of gas flow rate used in semiconductor process technology.
mass flow controller

SCE see *Short Channel Effects*.

Schottky barrier a potential barrier formed at the rectifying metal-semiconductor contact, i.e. at the contact between metal and semiconductor featuring different work functions.
metal-semiconductor contact, potential barrier, work function, Schottky barrier

Schottky clamped transistor *n-p-n* transistor with Schottky diode connecting base and collector; the role of Schottky diode is to drive excess minority carriers from the transistor base during switching, and hence, to reduce storage time.
Schottky diode, bipolar transistor

Schottky contact see *Schottky diode*.

Schottky defect a point defect in crystalline solid; vacancy.
crystal defects, point defect

Schottky diode semiconductor diode in which metal-semiconductor contact is used to form a potential barrier; potential barrier heights varies with bias voltage accounting for the diode-like (rectifying) current-voltage characteristics of the device.
diode, Schottky barrier

Schottky effect lowering of the potential barrier at the metal-semiconductor contact due to the high electric field.
Schottky barrier

Schottky emission an effect observed in metal-semiconductor contacts; emission of electrons from semiconductor to metal, or vice versa, due to the Schottky effect.
Schottky effect

Schrödinger equation also known as Schrödinger wave equation; fundamental relationship describing changes of the quantum state with time; in semiconductor physics accounts for the wave-like properties of an electron.

SCP, Surface Charge Profiling a non-contact electrical characterization methods of semiconductors; based on Surface Photovoltage (SPV) effect; allows determination of the density of surface charge and near-surface dopant concentration in semiconductors.
Surface Photovoltage

SCR Silicon-Controlled Rectifier, or thyristor; a device commonly used in high-power switching applications.
thyristor

screw dislocation a line defect; a slip of the part of the crystal with dislocation line moving perpendicular to the direction of stress in the lattice.
crystal defects, line defect, edge defect

scribing a process introducing stress along crystallographic planes in the single-crystal wafer; carried out by applying pressure to the diamond tipped scribing tool for the purpose of separating wafer into chips by crystal cleaving.
dicing, cleaving

scribing lines bare areas on the surface of the wafer in between the chips; form a network of lines along which wafer is scribed then cleaved into individual chips (dies).
scribing

scrubbing a process of removal of heavy contaminants from the wafer surface by means of mechanical interactions; used primarily after chemical-mechanical planarization (CMP) operations.
brush scrubbing, megasonic scrubbing, post-CMP cleaning

SCS Silicon-Controlled Switch.

SDHT Selectively Doped Heterojunction Transistor.

SE see *spectroscopic ellipsometry*.

Secco etch a solution of $K_2Cr_2O_7$ in water mixed with HF; an etchant used to reveal defects in single-crystal silicon.
Sirtle etch

Secondary Ion Mass Spectroscopy, SIMS very sensitive method to study chemical composition of solids; broadly used in semiconductor R&D and process diagnostics; atoms sputtered off the surface are identified by determination of their mass (mass spectroscopy); allows depth profiling; conventional dynamic SIMS features high sputtering rate while static SIMS uses very low sputtering rate allowing better sensitivity to the chemical characteristics of the surface; TOF-SIMS is used to identify species adsorbed at the surface.
Time-of-Flight SIMS, depth profiling

seed crystal a piece of single crystal material used to set a pattern for the growth of the crystal in which this pattern is reproduced.
Czochralski crystal growth, float-zone crystal growth

SEG Selective Epitaxial Growth; see *selective epitaxy*.

segregation coefficient, *m* at the same temperature the solubility of any given element in material *A* is in general not the same as in material *B*; *m* = the ratio of equilibrium concentration of an element in material A and equilibrium concentration of the same element in material B.
boron penetration, zone refining

SEL *see Surface Emitting Laser.*

Selective Area Chemical Vapor Deposition, SACVD CVD process which deposits thin film material in selected areas on the wafer surface only; selectivity of deposition is controlled by chemical composition of the surface which can be locally altered.
CVD, bottom-up processing

selective deposition see *Selective Area Chemical Vapor Deposition*; also deposition that can be limited to selected areas on the wafer by using masks.
mask

selective doping process introducing dopants to semiconductor in the selected areas only which are defined by the pattern created in the

195

masking material; as opposed to the introduction of dopants into the entire volume of semiconductor material during the crystal growth processes.
ion implantation, diffusion

selective epitaxy *(i)* epitaxial growth on the substrate which is only partially a single-crystal material; for instance in the case of single crystal Si partially covered with oxide, Si will grow epitaxially only (selectively) on the surface of a single-crystal Si; *(ii)* selective epitaxy is accomplished by the alteration of the chemical composition of the surface promoting epitaxial growth in selected areas only.
epitaxy, SACVD, bottom-up processing

selective etching the etching process in which etching medium reacts with one material on the wafer surface and do not react with others; e.g. $HF:H_2O$ solution is etching SiO_2 very rapidly while not etching silicon nitride Si_3N_4 and silicon.
non-selective etching, etch selectivity

selective exposure foundation of the pattern definition using lithographic processes; only selected areas of the resist are exposed to radiation; accomplished either by using masks or by localizing exposure using finely focused beam.
e-beam lithography, photolithography, mask, resist, direct-write lithography

SELED see *Surface Emitting Light Emitting Diode*.
EELED

self-assembled monolayer, SAM material which takes on desired shape and desired chemical composition due to its inherent characteristics and properly functionalized surface of the substrate on which it is formed.
bottom-up processing, surface functionalization

SEM see *Scanning Electron Microscopy*.

semi-insulating semiconductor semiconductor material featuring very high resistivity; only undoped semiconductors with very low intrinsic carrier concentration (i.e. featuring wide energy gap) can display semi-insulating characteristics; e.g. GaAs with intrinsic carrier concentration

$\sim 10^6$ cm^{-3} can be semi-insulating while Si with intrinsic carrier concentration $\sim 10^{10}$ cm^{-3} cannot; semi-insulating semiconductors allow better electrical isolation between adjacent devices.
intrinsic semiconductor

semiconductors solid-state materials in which, in contrast to metals and insulators: *(i)* electrical conductivity can be controlled by orders of magnitude by adding small amounts of alien elements (dopants), *(ii)* electrical conductivity can be controlled not only by negatively charged electrons, but also by positively charged holes, and *(iii)* electrical conductivity is sensitive to temperature, illumination, electric field, and magnetic field; in terms of chemical composition inorganic and organic semiconductors are distinguished; the former include elements from the group IV of the periodic table (elemental semiconductors) and a wide range of compound semiconductors synthetized using elements from the groups II to VI of the periodic table; used to fabricate electronic and photonic devices which define the progress of our technical civilization.
elemental semiconductors, compound semiconductors, inorganic semiconductors, organic semiconductors, semiconductor device

semiconductor device a very broad term; essentially any homogenous or multilayer piece of semiconductor material which is processed in such way that it can performed in the controlled fashion predetermined electronic (e.g. diode, transistor, monolithic integrated circuit) and/or photonic (e.g. LED, laser) functions.
electronic device, photonic device, discrete device, integrated device

semiconductor diode a diode fabricated using semiconductor materials; to obtain diode-like characteristics a potential barrier must be created within the semiconductor structure; implemented using *p-n* junction, or metal-semiconductor contact; height of the built-in potential barrier is controlled by the applied voltage; potential barrier at the *p-n* junction makes a *p-n* junction diode while the potential barrier at the metal-semiconductor junction makes a Schottky diode; along with a transistor, diode is the most important device configuration in semiconductor engineering; in addition to electronic functions, also used to fabricate LEDs (Light Emitting Diodes), lasers, and solar cells.
diode, p-n junction, metal-semiconductor contact, LED, semiconductor laser, solar cells

semiconductor engineering an engineering effort focused on the construction of functional devices and circuits using semiconductor materials.
semiconductors, semiconductor device, semiconductor process

semiconductor equipment a set of highly engineered tools used to fabricate semiconductor devices both discrete and integrated; equipment needed to implement semiconductor processes.
discrete device, integrated device, semiconductor process

semiconductor laser light emitting semiconductor diode which in contrast to conventional LED generates very sharp emission lines and allows modulation bandwidth in a gigahertz range; semiconductor laser is a complex multilayer device formed using direct-bandgap semiconductors primarily AIII-BV (e.g. GaAs, GaAlAs, GaN, etc.); its high performance is due to the enhancement of the recombination rate (stimulated emission) and presence of the optical cavity in the device structure.
LASER, LED

semiconductor lighting the use of LEDs in everyday lighting applications; LED bulbs replaced highly inefficient incandescent bulbs and less efficient fluorescent bulbs in all applications which use electric bulbs for lighting; LED bulbs use GaN-based heterojunctions and include phosphors to adjust spectrum of emitted light.
LED, phosphor

semiconductor manufacturing a distinct manufacturing technology featuring the highest level of complexity; uses the highest precision tools producing nanometer scale features in ultra-clean process environment.
manufacturing yield, semiconductor process

semiconductor materials see *semiconductors*.

semiconductor process term refers to the set of processes involved in the manufacture of semiconductor devices.
semiconductors, semiconductor engineering, semiconductor manufacturing

semiconductor sensors devices which change their electrical characteristics in response to external influences; e.g. MOSFET can be configured

such that its electrical characteristics will change in response to the change in the chemical composition of the ambient, gaseous or liquid; on the other hand a wide array of MEMS based sensors will respond to the physical forces acting upon them.
MOSFET, MEMS

series resistance the resistance of the selected part of the semiconductor device in the direction of current flow; typically resistance associated with limited conductivity of semiconductor itself, e.g. resistance of the bulk of semiconductor wafer in the case current flows from the front to the back surface of the wafer, or resistance of the metal contact to semiconductor.
ohmic contact, raised source-drain

shadow mask see *mask, mechanical.*

shadowing effect the effect preventing conformal coating of surface features during physical vapor deposition; e.g. evaporated metal reaching the surface under certain angle will not uniformly coat vertical walls and the bottom of the trench.
conformal coating, evaporation

shallow junction term typically refers to a depth of the source and drain regions in advanced CMOS; scaling rules require continued reduction of the junction depth; can be as shallow as 10 nm.
scaling rules

shallow trench isolation, STI an isolation scheme used in advanced integrated circuits, e.g. in CMOS ICs where it is used instead of the LOCOS isolation.
LOCOS, trench isolation, junction isolation

sheet resistance a measure of resistance of the very thin doped regions; expressed in Ω/square; commonly used to determine the outcome of semiconductor doping operations.

Shockley-Read-Hall, SRH, theory Shockley-Read-Hall theory of recombination; assumes presence of defects in semiconductor which are acting as recombination centers; in practice, quite often defect-related recombination rather than valence-to-conduction band recombination determines lifetime of excess charge carriers.
recombination, excess carriers

short-channel effects undesired physical effects (e.g. hot electron generation) occurring in the scaled down channel of a MOSFET; special measures must be taken to prevent short-channel effects; certain scaling rules must be followed while scaling down MOSFETs geometry.
channel, DIBL, hot electron, scaling rules

short-circuit current, I_{sc} a current of illuminated solar cell with shorted output.
open-circuit voltage, solar cell

SiC see *silicon carbide*.

sidewall vertical part (wall) of the trench; to assure it remains perpendicular to the wafer surface, sidewall during trench etching is covered with a layer of polymer.
anisotropic etching, Bosch process, sidewall passivation

sidewall passivation a process preventing lateral etching, i.e. assuring anisotropy of the etching process, during prolonged deep etches; accomplished by covering sidewalls with a layer polymer.
sidewall

sidewall roughness undesired roughness of the trench sidewall introduced during etching.
roughness

SiGe see *silicon germanium*.

silane, SiH_4 gaseous compound; the most common source of silicon in chemical vapor deposition processes of Si, SiO_2 and Si_3N_4; thermally decomposes at about 900 °C; toxic and explosive in contact with oxygen.
chemical vapor deposition

SILC Stress Induced Leakage Current; a test designed to study reliability of thin oxides in MOS gates; density of leakage current is measured following electric field stress of the oxide and compared with its density measured before the stress; useful in characterization of ultra-thin oxides in which "hard" breakdown does not occur due to the excessive tunneling across the oxide.
GOI, hard breakdown

silicene 2-dimensional (2D), one-atom thick silicon; a single-layer of hexagonally bonded silicon atoms displaying attractive physical properties; unlike graphene, silicene features energy gap; promising because of its obvious compatibility with silicon fabrication process technology.
graphene, molybdenum disulfide, phosphorene, stanene

silicidation an anneal (sintering) process resulting in the formation of metal-Si alloy (silicide) to act as a contact; e.g. Ti deposited on Si forms $TiSi_2$ silicide as the result of silicidation.
silicide, sintering, RTA, "salicide" process

silicide an alloy of silicon and metal; contact materials in silicon device manufacturing; e.g. $TiSi_2$, $CoSi_2$, $NiSi$; combines advantageous features of metal contacts (e.g. significantly lower resistivity than poly-Si) and poly-Si contacts (e.g. no electromigration).
metal-semiconductor contact, electromigration

silicon, Si element from group IV of the periodic table displaying excellent semiconductor properties; second the most abundant element in the Earth crust; energy gap E_g = 1.12 eV, indirect; crystal structure - diamond, mobility of electrons and holes at 300 K: 1450 and 500 cm^2/Vs respectively; the most common semiconductor; features excellent compatibility with broadly defined needs of semiconductor device manufacturing technology; used as a single-crystal, polycrystalline, multicrystalline, and amorphous material in the manufacture of logic and memory nanochips, analog ICs, myriad of discrete devices, thin-film transistors, solar cells, etc.; outstanding characteristic: native oxide of silicon, SiO_2, is a very high quality insulator; Si features excellent mechanical properties allowing critically important MEMS technology; unlike no other semiconductor single-crystal Si can be processed into wafers as large as 450 mm in diameter; forms compound semiconductors with elements form the group IV of the periodic table: silicon carbide (SiC) and silicon germanium (SiGe).
amorphous silicon, polycrystalline silicon, silicon carbide, -germanium

silicon active layer see *active silicon layer*.

silicon carbide, SiC group IV-IV semiconductor compound; also known as carborundum; can be obtained in several polytypes incuding hexagonal in the form of either 4H or 6H (α-SiC) and cubic 3C (β-SiC);

parameters vary depending on polytype; energy gap E_g = 2.9 - 3.05 eV (wide-bandgap semiconductor), indirect; carrier mobility at 300 K: electrons ~500 - 900 cm^2/Vs, holes ~20 - 50 cm^2/Vs; higher than Si and GaAs electron saturation velocity; thermal conductivity 3 W/cmK (two times higher than Si); excellent semiconductor in high-power, high-temperature device applications; difficult and expensive to fabricate in the form of single-crystal wafers featuring low defect density; also used as a substrate for GaN.
wide-bandgap semiconductor, saturation velocity, AIV-BIV

silicon dioxide, SiO₂ silica, glass; in a single-crystal form: quartz; native oxide of silicon and at the same time an excellent insulator; features energy gap E_g ~ 8eV; dielectric strength 5 - 15 x 10^6 V/cm depending on thickness and dielectric constant k = 3.9, the most common insulator in semiconductor device technology (where it is used only as an amorphous material), particularly in silicon MOS/CMOS where it is used as a gate oxide; formed by thermal oxidation on Si and by CVD on Si and other substrates; prone to contamination with alkali ions (sodium) and not resistant to high energy radiation.
gate oxide, dielectric strength, sodium, radiation hardness, thermal oxidation

silicon dopants elements from group III of the periodic table acting as acceptors (boron, B) and from group V of the periodic table acting as donors (antimony, Sb, arsenic, As, and phosphorous, P).
acceptor, donor

silicon etch wet chemical formulation designed to chemically polish and/or etch silicon; also used to delineated crystallographic defects through preferential etching.
CP-4, Secco etch, Sirtl etch

silicon germanium, SiGe group IV-IV semiconductor compound; in contrast to SiC does not exist in the define lattice but rather as a mixed crystal; an alloy featuring energy gap narrower than Si and wider than Ge and electron mobility higher than Si; energy gap and mobility varies as a function of Ge concentration in SiGe; single-crystal SiGe on single-crystal Si features strained lattice; epitaxially deposited to form base in HBT; in CMOS technology SiGe is used to introduce strain in the transistor's channel.
strained lattice, strained silicon, strain, CMOS, HBT, SiC

silicon nanowire, SiNW micrometer-scale in length and tens of nanometer in diameter, i.e. one-dimensional (1D) piece of typically crystalline silicon; single SiNW can be used as a channel in MOSFETs; interconnected nonowires form a circuit; ultra-small geometry of nanowire devices is defined by the geometry of the wire itself and not by the conventional method of pattern definition; also promising in several other application including energy conversion.
nanowire, nanotube

silicon nitride, Si₃N₄ dense, chemically resistant insulator (energy gap $E_g \sim 5eV$); dielectric constant $k \sim 6\text{-}7$; excellent mask (barrier) against oxidation of Si; commonly used in silicon integrated circuit manufacturing in LOCOS processes and as an etchstop; can be used as a gate dielectric, but not in Si MOSFETs due to inferior interface with silicon; used in double-layer gate dielectric memory cells (MNOS); properties depend on CVD chemistries used.
LOCOS, etchstop, MNOS

Silicon-on-Insulator, SOI basically a silicon wafer with a thin layer of oxide (SiO_2) buried in it (buried oxide, BOX) at certain distance (varied from several μm to < 10 nm) from the surface; devices are built into a layer of silicon (active layer) on top of the buried oxide; SOI substrates *(i)* provide superior isolation between adjacent transistors in an integrated circuit as compared to devices built into bulk wafers, *(ii)* eliminate "latch-up" in CMOS devices, *(iii)* improve performance of the SOI-built transistors due to reduced parasitic capacitances, *(iv)* are used in MEMS manufacturing in which case buried SiO_2 acts as a sacrificial material; fabricated by means of wafer bonding or SIMOX methods; in variations of SOI other than buried SiO_2 materials can be used as an insulator; e.g. SOS and SOAN.
bonded SOI, SIMOX, SOAN, SOS, ETSOI, buried oxide, fully depleted SOI, latch-up

Silicon-on-Sapphire, SOS special case of SOI where sapphire (single-crystal Al_2O_3) is an insulator (substrate) and active single crystal Si layer is formed on top of it by means of solid-phase epitaxy; developed mainly for very high-frequency analog IC applications where the fact that active Si layer is supported by homogenous bulk insulator (rather than by silicon underneath buried oxide) matters.
sapphire, SOI, solid phase epitaxy

silicon oxidation term commonly refers to the process of thermal oxidation which forms a thin layer of SiO_2 on silicon surface.
thermal oxidation

silicon oxide see *silicon dioxide.*

silicon oxynitride SiO_xN_y; mixture of silicon oxide and silicon nitride phase; can be formed by silicon oxidation in N_2O or NO or by anneal of SiO_2 in for instance NH_3.
nitrided oxide

silicon precursor silicon containing compound used as a source of silicon in deposition processes; e.g. SiH_4 or $SiCl_4$ in various silicon CVD processes.
silane, silicon tetrachloride

silicon-silicon dioxide, Si-SiO₂, system key element of the silicon MOS/CMOS gates; as such, very likely the most researched material system of all; formed by thermal oxidation of silicon; its most important part: an interface between Si and SiO_2 within which structural transition (from single-crystal Si to amorphous SiO_2) and chemical transition (from Si to SiO_2) take place; quality of Si-SiO₂ system has critical impact on the performance of silicon MOS/CMOS devices.
interface trap, oxide charge, silicon oxidation

silicon tetrachloride, SiCl₄ a gas, often used as a source of silicon; e.g. in epitaxial deposition of Si or deposition of poly-Si.
epitaxy, polycrystalline Si, silicon precursor

SiLK™ "silicon low-k"; a commercial name for the material used as an interlayer low-k dielectric (ILD); type of the silicon resin featuring dielectric constant $k = 2.65$.
inter-layer dielectric

SIMOX Separation by Implantation of Oxygen; a method used to fabricate SOI substrates; oxygen ions are implanted into Si substrate and form a buried oxide layer at certain distance from the surface; implantation of oxygen is followed with an anneal to remove implantation damage and to enforce reaction of oxygen with silicon (oxidation) in the implanted region.
silicon-on-insulators, bonded SOI

SIMS see *Secondary Ion Mass Spectroscopy.*

single-charge ion see *ion, single-charge.*

single-chip module package containing only one chip.
multichip module

single-crystal crystalline solid in which atoms are arranged following specific pattern throughout the entire piece of material; i.e., long-range order exists throughout; in general, single-crystal materials features superior electronic and photonic properties as compared to poly-crystalline and amorphous materials, but are more expensive; all high-end semiconductor electronic and photonic devices are fabricated using single-crystal substrates.
amorphous, polycrystalline

single-crystal growth see *Czochralski crystal growth, float-zone crystal growth.*

single damascene see *damascene process.*

single-wafer cleaning cleaning performed on one wafer only; reflects growing general trend toward single-wafer processing.
immersion cleaning, spin cleaning, single wafer process

single-wafer process only one wafer is processed at a time; opposite to batch process in which several wafers are processed simultaneously; tools are designed specifically for a single-wafer processing; becomes more common as wafer diameter increases.
batch process

single-walled nanotube, SWNT term refers to the carbon nanotube; basically a single layer of graphene rolled into a seamless cylinder and capped at both ends; one-dimensional (1D) material system; single nanometers in diameter; display unique electronic and mechanical properties; numerous applications possible.
carbon nanotube, multi-walled nanotube

sintering in semiconductor terminology an elevated temperature process forming an alloy of metal deposited on silicon with silicon; process forming silicide contacts in silicon devices.
silicides

sintering temperature temperature at which metal forms an alloy with silicon; lower sintering temperature the better.
sintering, silicidation, silicide

SiNW see *silicon nanowire.*

SIP System in Package; all components of the electronic system are contained in a single package.

SIP Single In-Line Package; single row of pins is located along one edge of the package.
dual in-line package

SIPOS Semi-Insulating Polycrystalline Silicon film; oxygen or nitrogen is added to silicon during growth; used for junction passivation.

Sirtl etch a solution of CrO_3 in water mixed with HF; etchant used to reveal defects in single-crystal silicon.
CP-4, Secco etch

slicing term refers to the process of cutting of the single-crystal semi-conductor ingot into wafers; high precision diamond blades are used; see also *wafering.*
wafer, wafer fabrication, ingot

slight etching the etching process aimed at the removal of just a few monolayers of the etched material.
atomic later etching

slow surface state surface state unable to exchange charge (charging and discharging) with a bulk of semiconductor in response to the fast changes of the bias voltage.
surface state, fast surface state

SLS see *strained layer superlattice.*

slurry a viscous liquid containing suspended abrasive component; used for lapping, polishing and grinding of solid surfaces; key element of CMP processes.
CMP, post-CMP cleaning

small-molecule organic semiconductors a class of organic semi-conductors; display properties superior to polymer based organic semiconductors, but typically require deposition by vacuum evaporation.
pentacene, organic semiconductor, polymer semiconductor

small signal electrical signal featuring amplitude low enough to consider system to which signal is applied linear.

Smart Cut® registered name (SOITEC, S.A.); the SOI wafer fabrication process based on wafer bonding; refers to the technique in which one wafer is cleaved after bonding along hydrogen implanted region and can be reused.
bonded SOI, SOI

SMIF Standard Mechanical Interface; a standard adopted by semi-conductor industry for a purpose of facilitating wafer transfer between "minienvironments".
minienvironment

SMIF pod, box a box (container) adapted to SMIF standards and used to transfer wafers between SMIF compatible process tools and "minienvironments" in a highly controlled ambient.
SMIF, minienvironment

SMOLED Small-Molecule Organic Light Emitting Diode; an OLED fabricated using small-molecule organic semiconductor.
OLED, small-molecule organic semiconductor

SMT see *Surface Mount Technology.*

SNOM Scanning Near-Field Optical Microscopy; see *NSOM.*

SOAN Silicon-On-Aluminum Nitride (AlN); type of SOI substrate in which AlN is a buried dielectric instead of SiO_2.
Silicon-on-Insulator

SOC see *System on Chip.*

SOD see *Spin-On Dielectric.*

sodium, Na an alkaline metal; one of the most common element in the environment; sodium ions Na^+ are harmful contaminants of SiO_2 in silicon processing; Na^+ ion can move in SiO_2 under the influence of electric field resulting in the instabilities of characteristics of MOS based devices; Na^+ contamination needs to be prevented in Si manufacturing environment.
mobile charge

soft bake a low temperature ($< 150\ °C$) bake (thermal treatment) to which wafer is subjected after deposition of photoresist; aimed at volatilization of solvents in the resist and its solidification; typically carried out using hot plate in ambient air.
hard bake, photoresist

soft breakdown excessive leakage current in the gate oxide but no irreversible hard breakdown of the oxide; a leakage current in MOS structure exceeds predetermined threshold value without casing permanent damage to the oxide; can be reversed by reducing field in the oxide.
breakdown, hard breakdown, reversible breakdown

soft lithography a block printing using elastometric (rubber) stamps and polymer ink; allows pattern definition in semiconductor applications where conventional photolithographic process and subsequent etching cannot be used, e.g. in organic semiconductor device manufacturing.
pattern, definition, organic semiconductor

SOG see *Spin-on Glass*.

SOI see *Silicon-on-Insulator*.

SOI wafer silicon wafer fabricated using Silicon-on-Insulator technology; thickness of silicon active layer varies from below 10 nm to several μm depending on application.
silicon on Insulator, wafer, epitaxial extension, engineered wafer

solar cell a two-terminal semiconductor device which converts solar light into electric signal; requires presence of the potential barrier within semiconductor which is typically accomplished by the formation of a *p-n* junction; implemented using various semiconductor materials (inorganic and organic) at various levels of structural complexity; choices are based

208

on the trade-off between cell's cost and efficiency; efficiency ~1% for low-cost, simple and over 45% for high-cost, complex solar cells respectively.
photovoltaic effect, fill factor, efficiency, excitonic-, organic solar cells

solid-phase crystallization, SPC process during which adequately executed heat treatments cause conversion of the amorphous phase in the solid into crystalline phase.

solid-phase epitaxy, SPE process during which, as a result of a heat treatment (often in properly selected gas ambient), an amorphized phase in the solid in contact with a single-crystal phase, re-crystallizes into a single-crystal.
amorphous material, epitaxy, Silicon-on-Sapphire

solid-solubility limit the number of atoms of an element A that can be incorporated into the solid B at any given temperature; an upper limit on the concentration of dopant atoms that can be introduced into semiconductor lattice at given temperature.
doping

solid-state light source, SSLS the terms refers to the LEDs used for lighting purposes.
white LED, Light Emitting Diode, semiconductor lighting

SOM Sulphuric (acid)-Ozone Mixture; H_2SO_4/O_3; a common cleaning solution designed to remove organic contaminants from the wafer surface; as more economical often replaces SPM cleaning mixture.
cleaning, organic contaminant, SPM

sonic wave a wave generated in the liquid chemical/water by megasonic agitation during wet cleaning operations; used to increase efficiency of particle removal from the wafer surface.
megasonic cleaning

SONOS Silicon-Oxide-Nitride-Oxide-Silicon; a non-volatile memory cell; features longer data retention, faster read time, and increased number of read-write cycles as compared to other floating gate memory technologies.
floating gate memory

SOS see *Silicon-On-Sapphire*.

source one of three terminals in the Field Effect Transistors; heavily doped, shallow region in semiconductor substrate from which current flows into the channel.
drain, JFET, MESFET, MOSFET

source-drain engineering changes in the configuration (depth in the Si substrate and heights above the Si surface) and in the doping level of source and drain regions in MOSFET/CMOS; needed to accommodate challenges related to device scaling.
raised source-drain, scaling

space charge see *space charge region*.

space-charge region a depletion region in semiconductor in which, due to the electric field present, there are no free charge carriers; result of the non-uniformity of potential distribution across semiconductor; e.g. potential barrier at the surface of semiconductor.
MOS, p-n junction, Schottky diode

spacer oxide deposited by CVD to isolate gate contact from source and drain contacts in a MOSFET/CMOS; also passivates sidewalls of the gate stack.
low-temperature oxide

SPC Statistical Process Control.

SPC see *Solid-Phase Crystallization*.

SPE see *Solid-Phase Epitaxy*.

spectroscopic ellipsometry, SE ellipsometry which uses more than one wavelength of light and allows variation of an angle of incidence; measures not only film thickness, but also provides information on select chemical/physical characteristics of the film(s); very useful in semiconductor process development, monitoring, and diagnostics.
ellipsometry

SPICE Simulation Program with Integrated Circuit Emphasis; simulator used to model electrical circuits at the transistor level, developed at the University of California at Berkeley.

spiking uncontrolled penetration of semiconductor substrate by contact metal; problem with Al in contact with silicon; may short ultra-shallow *p-n* junction underneath the contact causing its permanent failure.
barrier metal

spin see *electron spin.*

spin cleaning so-called "dynamic cleaning"; wafer rotated at high *rpms* is subjected to interactions with cleaning solutions and DI water; a single wafer process.
immersion cleaning, spray cleaning, single wafer process

spin drying removal of water from the surface of the wafer by centrifugal forces during fast spinning of the wafer; performance is inferior to IPA-based drying techniques.
IPA drying, Marangoni drying

spin-on deposition, spin coating process used to coat the wafer with material which is originally in the form of viscous liquid; liquid is dispensed onto the wafer surface in predetermined amount and the wafer is rapidly rotated (1500 - 7000 rpm); during spinning liquid is uniformly distributed on the surface by centrifugal forces; material is then solidified by a low-temperature (typically below 200 °C) curing; in semiconductor processing commonly used to deposit photoresist; not suitable for depositions on oddly shaped, heavy substrates.
mist deposition, Physical Liquid Deposition, spray deposition, inkjet printing

spin-on dielectric, SOD a dielectric in the form of viscous liquid is deposited by spin-on process then solidified through thermal curing.
spin-on deposition

spin-on glass, SOG thin layer of SiO_2 (often doped) deposited on the wafer surface by the spin-on process then solidified through thermal curing.
spin-on deposition

spin rinse-dry, SRD spinning of the wafer in deionized water and then in pure nitrogen.
drying, rinsing, deionized water

spin transistor, spin FET uses spin polarization of electron to distinguish between "on" and "off" state.
spintronics, Field-Effect Transistor

spintronics scientific and technical domain focused on using electron's spin, not its charge, to implement switching action in electronic devices; idea is based on the inherent binary nature of electron's spin states; spins can only be aligned up or down.
electron spin, magnetic semiconductor, electronics, photonics

SPM cleaning solution; involves $H_2SO_4:H_2O_2$ mixture typically in 1:4 ratio; strongly oxidizing cleaning solution is used to remove organic materials/contaminants, including remaining photoresist, from the wafer surface; typically applied first in the post-photolithography cleaning sequence.
SOM

spontaneous emission photon generation in semiconductor device as a result of spontaneous, as opposed to stimulated, band-to-band recombination; an effect upon which operation of light emitting diodes (LEDs) is based.
LED, radiative recombination, stimulated emission. laser

spray cleaning wet cleaning batch process in which wafers are sprayed with cleaning chemicals and then rinsing water; alternative to immersion cleaning and spin cleaning.
immersion cleaning, spin cleaning

spray deposition liquid precursor, e.g. photoresist, is sprayed on the surface of the substrate; droplet size is in the range of 20 μm; works best for films thicker than about 1 μm; in contrast to spin-on process, spray deposition is independent of the substrate size and shape.
mist deposition, spin-on deposition, Physical Liquid Deposition

spreading resistance a resistance of semiconductor defined by the distribution of dopant atoms in the direction normal to its surface.
doping, resistivity

spreading resistance profiling determination of depth distribution (profile) of dopant atoms in semiconductor substrate and junction depth.
junction depth

sputtering, sputter deposition an additive process; bombardment of a solid (target) by high energy chemically inert ions (e.g. Ar$^+$) extracted from plasma; causes ejection of atoms from the target which are then re-deposited on the surface of the substrate located in the vicinity of the target; common method of Physical Vapor Deposition (PVD); can be used to deposit in the form of a thin-film essentially any solid.
additive process, physical vapor deposition, magnetron sputtering

sputtering, sputter etching a subtractive process; same as ion milling; bombardment of the target (wafer) by high energy ions (e.g. Ar$^+$) extracted from plasma for the purpose of material removal; physical etching through momentum transfer; highly anisotropic, highly non-selective etching method.
dry etching, ion milling, physical etching

sputtering target a source material during sputter deposition processes; typically in the form of the disc which inside the vacuum chamber is exposed to the bombarding ions.
sputtering

sputtering yield efficiency of the sputtering process; sputtering yield is different for different materials.
sputtering

SPV see *Surface Photo-Voltage.*

SRAM see *Static Random Access Memory.*

SRD see *Spin Rinse-Dry.*

SRH see *Shockley-Read-Hall theory.*

SRH recombination see *Shockley-Read-Hall theory.*

SRO Stress Relieve Oxide; oxide film formed on the wafer surface to reduce stress in the wafer surface region.

SSI Small Scale Integration.
VLSI

SSLS see *solid-state light source.*

sSOI strained SOI, Silicon-on Insulator substrate with a strain built into the lattice of an active Si layer.
SOI, strain, active Si layer

stacking fault a defect in the single-crystal material; planar defect; disruption of the stacking sequence of crystallographic planes; frequently observed during epitaxial growth.
crystal defects, epitaxy

Staebler-Wronski effect degradation of electrical output of amorphous silicon solar cells as the result of prolonged exposure to sunlight.
hydrogenated a-Si, solar cell

staggered gap a lineup of the edges of conduction and valance bands in semiconductors forming an abrupt heterojunction such that the edges of both conduction and valence bands in one semiconductor forming heterojunction are below edges of the respective bands in the other.
band-edge lineup, abrupt heterojunction, broken gap, straddling gap

stanine a single atom thick (2D material) tin.
graphene, molybdenum disulfide, silicene, phosphorene

standing wave effect undesired effect of standing wave formation in the photoresist during exposure; caused by the reflection of light from the surface upon which photoresist is deposited; controlled by deposition of an anti-reflective coating under the photoresist and post-exposure low temperature anneal.
bottom anti-reflective coating

static charge electric charge that is inadvertently accumulated on the surface of semiconductor wafer during wafer handling and processing; must be minimized to prevent its harmful effect of device structures at various stages of manufacturing process, e.g. may cause breakdown of the very thin oxides; also, static charge accumulating within the clean-room infrastructure is a challenge to semiconductor manufacturing processes.
clean-room, semiconductor manufacturing

Static Random Access Memory, SRAM read-write memory cell; in contrast to DRAM, data is stored permanently using combination of transistors (no capacitors) and does not need to be refreshed periodically, hence "static"; SRAM cell is significantly larger than DRAM cell.
DRAM

steam oxidation see *wet oxidation.*

steep-slope transistor a MOSFET featuring steep slope of the drain current – gate voltage characteristics in sub-threshold region; results in better switching performance of the transistor.
sub-threshold region, MOSFET, Tunnel FET

STEM see *Scanning Transmission Electron Microscopy.*

stencil mask see *mask, mechanical.*

step-and-repeat projection resist exposure mode in lithographic processes in which the same pattern is projected on the surface of the wafer in the series of consecutive exposures instead of the exposure of the entire wafer.
full-field exposure, stepper, projection printing

step coverage ability of thin-film deposition method to assure uniform coating over the geometrical features (steps) existing on the surface of the wafer.
conformal coating

stepper a photoresist exposure tool commonly used in photolithograhy; works using projection printing; in contrast to full-field exposure tools, stepper exposes through the remotely located reticle and lens system only part of the photresist covered wafer; process is repeated ("step-and-repeat") as many time as needed to expose the entire wafer.
projection printing, reticle, step-and-repeat, full-field exposure

STI see *Shallow Trench Isolation.*

stimulated emission process that increases rate of photon generation in semiconductor; basis of the operation of semiconductor lasers.
laser, spontaneous emission.

STM   see *Scanning Tunneling Microscopy.*

STO   see *strontium titanate.*

storage capacitor   e.g. key element of DRAM cell.
DRAM

storage time   time needed for the minority carrier concentration in the vicinity of *p-n* junction to be reduced to zero after junction is switched from forward to reverse bias; effect which has an adverse impact on the switching characteristics of bipolar devices.
Schottky clamped transistor

stradding gap   a lineup of the edges of conduction and valance bands in two different semiconductors forming an abrupt heterojunction such that the edge of the conduction band in one semiconductor forming hetero-junction is below the edge of the conduction band in the other semiconductor while the edge of the valence band is higher than in the other one.
band-edge lineup, abrupt heterojunction, broken gap, staggered gap

straggle   standard deviation of the Gaussian distribution of implanted ions in the direction of ions motion.
ion implantation, transverse straggle

strain   in semiconductor terminology a deformation of crystal lattice due to the stress; strain represents relative change in bonds configuration in the crystal and is unit-less; strain in the lattice increases mobility of charge carriers, and hence, is often introduced to the crystal on purpose to favorably modify charge carrier transport.
stress, strained layer

strain, biaxial   see *biaxial strain.*

strain, global   see *biaxial strain.*

strain, uniaxial   see *uniaxial strain.*

strained layer   an ultra-thin layer of single-crystal material epitaxially grown on the substrate featuring different lattice constant; pseudo-morphic material; difference in lattice constant causes lattice of the

deposited film to be strained; effective mass of electrons in strained material is reduced, and hence, their mobility is higher than in the relaxed material.
lattice constant, lattice mismatched structure, pseudomorphic material

strained layer quantum-well device a quantum-well device comprising several strained layers.
quantum-well, strained layer

strained layer supperlattice, SLS a structure comprising of several epitaxial layers of lattice mismatched materials thin enough to avoid formation of dislocations (pseudomorphic films); strain affects electronic properties of materials involved; by controlling thickness of each film and its chemical composition the SLS structures can be designed to perform variety of electronic and photonic functions in various device configurations.
lattice mismatch, pseudomorphic material, superlattice

strained silicon very thin layer of single-crystal silicon with built-in strain to increase mobility of electrons; allows manufacture of faster devices; strain in Si lattice can be accomplished by depositing ultra-thin layer of silicon on SiGe (different lattice constant).
strained film

stress represents a force per unit area; in semiconductor terminology typically refers to the stress at the interfaces between materials featuring different crystallographic structure and/or different thermal expansion coefficients resulting from different chemical composition; physical damage to the surface also introduce stress in the near-surface region.
compressive stress, tensile stress

stressor a material added to the single-crystal semiconductor to introduced stress in its crystal structure; typically accomplished by deposition of the lattice mismatched material.
lattice mismatched structure, pseudomorphic film

stripping process of material removal from the entire wafer surface; typically implies that the removal is not carried out for the pattering purpose, e.g. resist stripping in which case entire resist is removed following photolithography and etching.
etching

strong inversion an inversion in the sub-surface region of semiconductor in MOS devices gets to the point where surface potential reaches twice the value of Fermi potential in the bulk of semiconductor; in the strongly inverted region concentration of minority carriers exceeds dopant concentration.
Fermi potential, inversion, channel

strontium titenite, SrTiO₃ dielectric constant $k = 90 - 240$; depending on crystallographic structure and thickness displays either dielectric, or ferroelectric properties.
high-k dielectric, complex oxide

structural defect see *crystal defect*.

sub-collector contact a region in the collector of bipolar transistor immediately underneath collector contact; should be heavily doped to form low-resistance ohmic contact between collector (semiconductor) and collector contact (metal); with very heavy doping of sub-collector region the potential barrier at the contact-semiconductor interface is so thin that carriers can tunnel through it.
bipolar transistor, tunneling barrier

suboxide oxide featuring less oxygen atoms than needed to fully oxidize Si atom; SiO_x with x varying from 0.5 to 1.5; e.g. $SiO_{1.5}$ features three oxygens instead of four needed to form stoichiometric SiO_2.
silicon dioxide

substitutional defect an alien atom in the crystal lattice substituting for the host atom in the lattice position.
interstitial defect, point defect

substitutional diffusion diffusion mechanism in which diffusant substitutes for the host atoms by displacing them from their lattice sites; high activation energy process; substitutional diffusants (e.g. all *p*- and *n*-type dopants in silicon) feature low diffusion coefficient.
diffusant, diffusion, interstitial diffusion

substrate injection during the constant-current stress of the MOS gate stacks (oxide reliability test) electrons are injected into the oxide from the Si substrate (negative bias on the substrate).
constant-current stress, gate injection

subthreshold conduction the current conduction in the MOSFET biased below the threshold inversion point.
threshold inversion point

subthreshold region a region on MOSFET output characteristics just below threshold voltage.
threshold voltage

subthreshold slope a slope of the MOSFET's output current-voltage characteristics just below threshold voltage.
threshold voltage, subthreshold region, steep-slope transistor

subtracting process a process which removes top layers of the solid, essentially an etching process; as opposed to additive process.
additive processes

sulfur passivation passivation of the surfaces of selected III-V compounds with sulfur, e.g. GaAs and GaSb.
surface passivation, hydrogen passivation

supercritical cleaning cleaning technique which uses super-critical fluid as a carrier of a cleaning agent.
supercritical fluid

supercritical CO_2, SCCO2 a gas (CO_2) of choice for supercritical fluid cleaning applications; critical point (transition from gas to supercritical fluid) at 31 °C and 31 atm.
supercritical fluid

supercritical fluid a state of matter into which gases and liquids are converted at elevated temperature and high pressure; supercritical fluid has characteristics of both liquid and a gas; it lacks any surface tension while interacting with solid surfaces, and hence, can readily penetrates high aspect ratio geometrical features; it has very low viscosity and, like a liquid, easily dissolves large quantities of other chemicals; in semiconductor processing used in surface cleaning when penetration of very tight geometrical features is required; typically supercritical CO_2 is used as carrier of cleaning agents.
supercritical CO_2

superlattice semiconductor structure comprising of several ultra-thin layers (typically 1.0 - 2.0 nm thick) engineered to obtain specific electronic and photonic properties; slight modifications of chemical composition of each layer result in slight variations of the energy bandgap from layer to layer (bandgap engineering) while the same crystallographic structure is maintained throughout the entire material system; fabrication of superlattices requires high-precision hetero-epitaxial deposition methods such as MBE and MOCVD; typically involves III-V semiconductors.
bandgap engineering, lattice matched structures, molecular beam epitaxy

SUPREM Stanford University Process Emulation Module; software allowing simulation of key processes in semiconductor manufacturing.

surface in general, the outermost layer (with no thickness) of a solid or a liquid body; in semiconductor terminology an abrupt discontinuity of material structure with very specific, and different from the bulk, physical and chemical properties, e.g. as opposed to the bulk is not electrically neutral.
bulk

surface amorphization conversion of the surface and near-surface region of a single-crystal semiconductor into amorphous phase; accomplished either by physical (e.g. high energy ion bombardment), or chemical (adequately formulated wet etch chemistries) means.
amorphous material

surface analysis a process aimed at the determination of chemical composition (e.g. XPS) and/or physical structure (e.g. RHEED) of solid surfaces.
RHEED, XPS

surface charge electric charge present on semiconductor surface; originates from disrupted lattice structure and/or from interactions with process/storage ambient; surface charge alters distribution of charge in the near-surface region of semiconductor.
surface

surface charge analysis involves methods conceived to measure density of surface charge by means of non-contact electrical characterization

techniques; typically based on the Surface Photovoltage effect (SPV); one approach is to measures width of the space charge region at the semiconductor surface from which density of surface charge as well as surface recombination lifetime can be derived.
non-contact electrical characterization, surface charge, surface photo-voltage

surface cleaning see *cleaning*.

surface conditioning surface treatments aimed at establishing specific chemical characteristics of the semiconductor surface; for instance hydrogen termination of Si surface; often uses the same wet and dry techniques as those used in surface cleaning.
hydrogen termination, surface passivation

surface damage process related disruption of the crystallographic order at the surface of single-crystal semiconductor substrates; typically caused by surface interactions with high energy ions during dry etching and ion implantation.
ion implantation, reactive ion etching damage

surface dopant concentration dopant concentration in the region immediately adjacent to the surface of doped semiconductor; may be different than dopant concentration in the bulk of semiconductor; needs to be monitored, particularly in the case of epitaxial layers.
dopant, epitaxial layer, dopant deactivation

Surface Emitting Laser, SEL same as VCSEL; generated radiation is propagated in the direction normal to the surface; potentially superior to Edge Emitting Laser (EEL) providing strict requirements regarding device configuration are met; allows more efficient coupling to the optical fiber than EEL.
edge emitting laser, laser, VCSEL

Surface Emitting Light Emitting Diode, SELED LED designed such that the light generated emerges from the device in the direction normal to the junction plane, and hence, to the surface; allows more efficient coupling to the optical fiber than edge emitting LED (EELED).
edge emitting LED, LED

surface energy   a measure of the tendency of a surface to repel a liquid; surface energy quantifies the disruption of intermolecular bonds that occur when a surface is created; the surface energy may be seen as the excess energy at the surface of a material compared to the bulk; it is the sum of all the excess energies of the surface atoms; low surface energy refers to a surface's repelling tendency (hydrophobic); high surface energy refers to a surface's attracting tendency (hydrophilic); surface energy of liquids: surface tension.
surface, hydrophilic surface, hydrophobic surface

surface engineering   a set of processes devised to established desired physical and/or chemical condition of semiconductor surface.
surface conditioning, surface reconstruction

surface functionalization   surface processed in such way that it will interact in the desired fashion with an ambient to which it is exposed; can be applied selectively so that on only parts of the surface will acquire desired characteristics.
bottom-up processing, self-assembled monolayer

surface mount technology, SMT   a method used to connect packaged microchip to printed board; no through-holes in the board are required; package leads are soldered directly to the board surface.

surface orientation   defines crystallographic structure of semiconductor surface using Miller indices; crystallographic plane along which single-crystal semiconductor ingot is cut into wafers; e.g. (100) and (111).
crystal plane, Miller indices

surface passivation   termination of bonds on the semiconductor surface with elements assuring chemical stability of the surface; surface is rendered chemically "passive"; e.g., hydrogen termination of bonds on the silicon surface will prevent oxidation of the surface; oxidation of Si surface bonds will also passivate the surface.
hydrogen termination, oxidation

Surface Photovoltage, SPV   generation of surface potential in semi-conductors as a result of illumination with light featuring energy higher than energy gap of semiconductor; photo-generated electrons and holes are separated by the electric field in the space charge region at the surface.
photovoltaic effect, surface charge analysis

surface planarization a process performed on the patterned (uneven) surface of the wafer for the purpose of rendering it flat; see *planarization*.
planarization, Chemical-Mechanical Planarization

surface potential a measure of semiconductor surface departure from the state of electrical neutrality; results in the energy bands bending at its surface; quantitatively surface potential is a difference between intrinsic Fermi level at the surface and intrinsic Fermi level in the bulk, i.e. in undisturbed (electrically neutral) part of semiconductor.
barrier heights, intrinsic Fermi level

surface properties physical properties of the material at its surface; inherently different from the bulk properties, e.g. minority carrier lifetime at the surface is significantly lower than in the bulk of the same wafer.
bulk properties

surface recombination recombination of free charge carriers in semiconductor *via* the electrically active centers (defects) at its surface.
recombination, recombination site

surface recombination velocity a rate of surface recombination; higher in the case of highly defective semiconductor surfaces.
surface recombination, recombination site

surface reconstruction process that establishes crystallographic structure of the surface different than the structure in the bulk of the crystal, i.e. process by which, driven by the need to lower surface energy, atoms at the surface of the crystal assume different structure than that in the bulk; mechanism of reconstruction depends on crystal orientation, e.g. is different in the crystals featuring different surface orientation.
surface orientation

surface roughness disruption of the planarity of the semiconductor surface; three-dimensional surface morphology; measured (typically by AFM) as a difference between highest and deepest surface features; can be as low as about 0.05 nm for high quality Si wafers with epitaxial layers and as high as in μm range for low-quality, or damaged semiconductor surfaces; has a detrimental effect on device performance, e.g. lower breakdown of the gate oxide formed on the rough surface.
AFM, oxide breakdown

surface scattering a charge carrier scattering mechanism specific to the near-surface region; highly intensified carrier scattering as compared to the carrier moving in the bulk of semiconductor; a key reason for the carrier mobility in the MOSFET channel to be significantly lower than in the bulk of semiconductor.
scattering, mobility

surface states electrically active states resulting from the disruption of the periodicity of the lattice at the semiconductor surface; energetically represented by the discrete energy levels in the energy gap of semiconductor at the surface.
energy gap, surface

surface tension a surface energy of liquids.
surface energy

surround gate the gate in MOSFET comprising CMOS cell designed in such way that it surrounds the channel; the goal is to increase gate area of the MOSFETs forming CMOS cell without increasing area of the cell.
double gate, tri gate, MOSFET

SWAMI Sidewall Masked Isolation; complex, and hence, rarely used variation of LOCOS process in which "bird beak" is minimized.
"bird beak", LOCOS

SWCNT Single Walled Carbon NanoTube.

SWNT see *Single Walled NanoTube*.

SWP see *Single-Wafer Process*.

synchrotron a source of high intensity X-rays suitable for X-ray lithography; very large and very expensive installation not used in commercial chip manufacturing.
X-ray lithography, synchrotron radiation

synchrotron radiation high-intensity X-ray radiation needed to expose the resist in X-ray lithography; X-rays are produced as a result of electron beam bending under the condition of electron synchrotron resonance.
X-ray lithography

System-on-Chip, SOC   all components of a complex electronic system are integrated on a single chip.
chip

T

TAB   Tape Automated Bonding; chip packaging technology; one-step process used to connect chip to the package; package leads are formed on the flexible tape.
package

tandem cell   a multi-junction, multi-material (single-crystal) solar cell designed to absorb and convert into electricity as large portion of the solar spectrum as possible.
solar cell, heterojunction

tantalum, Ta   refractory metal; element in compounds, both conductive (e.g. TaN, $TaSi_2$) and insulating (Ta_2O_5), used in semiconductor processing.
R-MOSFET, barrier metal

tantalum carbide, TaC   Ta based conductor; possible MOS gate contact material in high-k dielectric MOS gate stacks.
gate contact, HKMG

tantalum nitride, TaN   a barrier metal used in Cu metallization scheme.
barrier metal, titanium nitride

tantalum pentoxide, Ta_2O_5   an oxide featuring dielectric constant $k \sim 25$; due to the lack of thermal stability with silicon cannot be formed on Si substrate without formation of an excessive interfacial SiO_x.
high-k dielectric, titanium dioxide, interfacial oxide

tantalum silicide, $TaSi_2$ contact (ohmic) material in Si technology; resistivity 35 - 55 $\mu\Omega$-cm; formed at the sintering temperature of 800 °C - 1000 °C.
silicide, ohmic contact, sintering

tapping mode   see *Atomic Force Microscopy.*

target source material used during sputter deposition; typically in the form of high purity disc installed in the sputtering system in such way that it is directly exposed to plasma.
sputtering

TBD see *Time to Breakdown.*

TCA trichloroethane; a liquid; organic solvent; its vapor is used as a source of chlorine in Si processing, e.g. during thermal oxidation to complex metallic contaminants potentially present on the Si surface.
thermal oxidation, metallic contaminants

TCE trichloroethylene; a liquid; organic solvent; its vapor is used as a source of chlorine in Si processing, serves similar perposes as TCA.
TCA

TCO Transparent Conductive Oxide (e.g. indium tin oxide or ZnO)
ITO, zinc oxide

TDDB see *Time-Dependent Dielectric Breakdown.*

TD-GC-MS Thermal Desorption Gas Chromatography Mass Spectroscopy; variation of GS-MC method; used to detect organic contaminants on semiconductor surface.
GC-MS, organic contaminant

technology generation in semiconductor terminology term used in reference to the classes of integrated circuits typically distinguished based on the transistor's gate length in nanometers; lower numbers (shorter gate) represent more advanced technology generations.
gate length

technology node a number in nanometers defining technology generation.
technology generation

TED see *Transient Enhanced Diffusion.*

TED Transferred-Electron Diode; microwave device also known as Gunn diode.
Gunn diode

telegraph noise see *Random Telegraph Noise.*

TEM see *Transmission Electron Microscopy.*

tensile stress stress resulting from the forces acting toward tearing material apart; compressive stress acts in the opposite direction; stress is a common occurrence in semiconductor structures at the interface between crystals featuring different lattice constants.
compressive stress, lattice mismatch

TEOS Tetraethyl Orthosilicate, $Si(OC_2H_5)_4$; gaseous compound commonly used in CVD of SiO_2 processes (so-called "TEOS Oxide"); good conformality of coating; relatively inert, liquid at room temperature; thermally decomposes at around 700 °C to form SiO_2; plasma enhancement lowers temperature of deposition to < 500 °C.
conformal coating, CVD, PECVD

TeraHertz Transistor transistor operating in the 10^{12} Hz (1,000 GHz) frequency range; requires SOI/SOS substrates and other modifications of its structure and material technology to operate in THz regime.
SOI SOS

ternary semiconductor semiconductor compound consisting of three elements; e.g. AlGaAs (III-III-V compound) or CdHgTe (II-II-VI compound); in the former case used in multilayer heterostructures for the purpose of bandgap engineering.
binary semiconductor, quaternary semiconductor, elemental semiconductor, compound semiconductor, bandgap engineering

TFET see *Tunnel FET.*

TFT see *Thin Film Transistor.*

thermal budget term defines total amount of thermal energy transferred to the wafer during elevated temperature operation; proportional to temperature and duration of the process; low thermal budget is desired in ultra-small IC manufacturing to prevent dopant redistribution; low thermal budget is possible even at a very high temperature if time of the process is very short (can be as short as few seconds); quantitatively proportional to Dt factor (D – diffusion coefficient, t – time of thermal treatment).
diffusion coefficient, dopant redistribution, RTA, RTP

thermal conductivity a measure of the solid ability to conduct heat; in semiconductor devices higher the thermal conductivity of the substrate material the better (higher efficiency of the removal of heat dissipated into substrate during device operation); unit: W/cmK; thermal conductivity of diamond is 22 W/cmK, as compared to 1.48 W/cmK for Si, 0.46 W/cmK for GaAs, and 4.9 W/cmK for 6HSiC.
diamond, critical temperature, heat management

thermal equilibrium condition in which the rate of generation of electrons and holes in semiconductor equals the rate of electrons and holes recombination.
generation, recombination

thermal evaporation see *filament evaporation* and *electron beam (e-beam) evaporation.*
PVD

thermal generation process generating free charge carriers in semiconductor by thermally stimulated band-to-band generation.
band-to-band generation

thermal noise distortion of electrical signal caused by thermal agitation of charge carriers in semiconductor.
noise, flicker noise

thermal oxidation growth of native oxide of the solid through oxidation of solid's surface at elevated temperature; results from thermally stimulated chemical reaction of host atoms with oxygen containing ambient; term commonly refers to thermal oxidation of silicon which results in a very high quality silicon dioxide, SiO_2, formed on its surface.
silicon dioxide, Deal-Grow model, dry oxidation, wet oxidation

thermal oxide oxide grown of the solid surface in the course of thermal oxidation; synonymous with thermal SiO_2 which is the result of thermal oxidation of Si; with an exception of silicon carbide, SiC, no other semiconductor forms device quality thermal oxide.
silicon dioxide, silicon carbide

thermal velocity an average velocity of charge carriers in semiconductor determined by the temperature of the lattice.
charge carriers

thermionic emission transport of electrons over the potential barrier at the metal-semiconductor interface or potential barrier created by the oxide in MOS structures.
potential barrier, metal-semiconductor contact

thick film material in the form of the film, but displaying basic physical properties (e.g. resistivity) which are not significantly different from the physical properties of the same material in the bulk form; typically, films in the thickness range from about 10 μm to 500 μm and above are referred to as thick; deposited mostly by screen printing.
bulk, thin-film, hybrid IC

thin film material in the form of the film thin enough to display physical properties (e.g. resistivity) distinctly different from the physical properties of the same material in the bulk form; difference results from the two-dimensional confinement of charge carriers in the thin film; in semiconductor technology typically thinner than about 0.5 μm (500 nm).
bulk, thick film

thin-film devices semiconductor devices fabricated using thin-film technology.
Thin-Film Transistor, thin-film solar cell

thin-film solar cell solar cell fabricated using thin-film non-crystalline semiconductors such as, most commonly, amorphous silicon (a-Si), but also cadmium telluride (CdTe) and organic semiconductors.
solar cell, amorphous silicon, cadmium telluride, organic solar cells

Thin-Film Transistor, TFT a Metal-Oxide-Semiconductor Field Effect Transistor implemented using thin-film technology; uses non-crystalline thin films of either inorganic (amorphous Si, ZnO) or organic (OTFT) semiconductors deposited on insulating substrates (e.g. glass); key element of active matrix liquid crystal display (LCD) technology; represents simplest configuration of device capable of implementing transistor action; can be formed of either rigid or flexible substrates.
MOSFET, organic TFT, bottom-gate TFT, top-gate TFT

three-dimensional (3D) integrated circuit instead of increasing an area of the chip built into a substrate wafer, chip is constructed in layers of single-crystal silicon stacked vertically and vertically interconnected

to form a functional circuit; the challenge is to form defect-free single-crystal Si layers for upper layers of the stack.
chip, vertical interconnects

three-dimensional (3D) integration (3D Packaging) term refers to IC packaging technology; prior to packaging fully processed planar chip is separated into functional sub-chips (blocks) which are thinned then stacked and vertically interconnected with Through Silicon Vias; 3-D integration improves performance (shorter interconnects) and reliability of an IC and allows its higher density.
packaging, vertical interconnect, via, wafer thinning, through silicon via

three-dimensional (3D) material term basically refers to the bulk material featuring the same physical properties in all three directions.
bulk, two-dimensional material

threshold adjustment control of the threshold voltage (V_T) of the MOSFET/CMOS by altering dopant concentration in the channel using ion implantation.
threshold voltage, channel, ion implantation

threshold inversion point an onset of inversion, the condition in which the inversion charge density equals semiconductor doping concentration; in other words a point at which channel is created in the MOSFET.
Inversion, channel, MOSFET

threshold voltage, V_T voltage applied to the gate of a field effect transistor (FET) that is necessary to open conductive channel between source and drain; in the case of MOSFET V_T is a voltage necessary for the inversion layer to be formed at the semiconductor surface.
channel, inversion, threshold inversion point

Through Silicon Via, TSV involves etching vias through the Si wafer to allow vertical wafer-to-wafer (chip-to-chip) interconnect scheme; compatible with 3D wafer packaging; typically requires wafer thinning; improves performance of the packaged chip by eliminating wire bonding and shortening interconnects lengths.
three-dimensional integration, wafer thinning, wire bonding

throughput in semiconductor terminology term typically refers to the number of wafers processed per hour by the process tool; high throughput tools/processes are desired.

thyristor a four-layer, three-junction (*p-n-p-n*) semiconductor device for high power switching; operates at very high voltage (~10 kV) and current (~5 kA) levels; also called silicon-controlled rectifier (SCR).
power device, p-n junction

Time-Dependent Dielectric Breakdown, TDDB technique used to evaluate reliability of gate oxides in MOS devices; breakdown results from the prolonged voltage stress; time needed to break stressed oxide is measured either under the Constant Voltage Stress (CVS), or Constant Current Stress (CCS).
oxide breakdown, constant-current, constant-voltage stress, charge-to-breakdown, Time-Zero Dielectric Breakdown

Time-of-Flight Secondary Ion Mass Spectroscopy, TOF-SIMS type of SIMS in which incident ion beam featuring ultra-low current is used; virtually non-destructive; information regarding chemical composition of very top surface of solids can be obtained; useful in detection of organic compounds adsorbed on the surface.
Secondary Ion Mass Spectroscopy

Time-to-Breakdown, TBD time it takes to break the oxide in the course of the constant current test; knowing TBD and constant current density set for the test allows determination of the charge-to-breakdown.
Time Dependent Dielectric Breakdown, charge to breakdown

Time-Zero Dielectric Breakdown technique used to evaluate reliability of gate oxides in MOS devices; oxide breakdown occurring without prolonged voltage stress i.e. under the condition of voltage stress rapidly increasing from zero to oxide breakdown.
oxide breakdown, ramp voltage breakdown, time-dependent dielectric breakdown

tin, Sn an element from group IV of the periodic table; in contrast to other group IV elements displaying strong semiconductor characteristics (C, Si, Ge), elemental bulk Sn is not an useful semiconductor; forms semiconductor compounds, e.g. GeSn; also promising in electronic applications as a single-atom thick (2D) material called stanene.
germanium tin, stanene

titanium, Ti a metal; element in compounds both conductive (e.g. TiN, TiSi$_2$) and insulating (TiO$_2$) used in semiconductor processing.
titanium nitride, titanium oxide

titanium dioxide, TiO$_2$ dielectric material which dielectric constant k varies from ~20 to ~85 depending on its structure and thickness; not thermally stable with silicon, and hence, not suitable as a high-k dielectric for MOS gates on silicon.
high-k dielectric, tantalum pentoxide

titanium nitride, TiN "tinitride"; conductor which resistivity increases with nitrogen content; used in silicon technology as a barrier separating silicon and metal contact (e.g. to control junction spiking); high melting point (2950 °C); commonly deposited by LPCVD.
barrier metal, spiking, LPCVD

titanium silicide, TiSi$_2$ contact (ohmic) material in Si technology; resistivity 13 - 16 μΩ-cm; formed at the sintering temperature of 700 °C - 900 °C.
silicide, ohmic contact, sintering temperature

titanium sublimation pump an ultra-high vacuum (~10^{-9} torr) pump; gas molecules are removed through chemical reactions with titanium sublimed onto pump walls.
high vacuum, turbomolecular pump

TOF-SIMS see *Time-of-Flight Secondary Ion Mass Spectroscopy.*

top-down process a way of building semiconductor structures using sequence of deposition-pattern definition-etching steps; as opposed to bottom-up processing.
planar process, bottom- up process

top-gate TFT a Thin-Film Transistor with a gate contact and gate dielectric on top of the deposited first thin-film semiconductor.
Thin-Film Transistor, bottom-gate TFT

torr unit of pressure commonly used in semiconductor technology; 1 standard atmosphere = 760 torr = 101325 Pa; 1 torr = 133.3 Pa.
pascal

Total Reflection X-Ray Fluorescence spectroscopy, TXRF a method of surface analysis; very effective in detecting metallic contaminants on the wafer surface; detection of the energy of photons emitted from atoms on the surface as the result of X-ray irradiation at the angle assuring total external reflectance.
metallic contaminant, X-ray fluorescence

transconductance, g_m parameter defining conditions for current flow in the channel of Field Effect Transistors (FETs); the derivative of the output current over input voltage.
channel

transient enhanced diffusion, TED transient effect observed in implanted regions of semiconductor during post-implantation annealing; due to the crystal damage resulting from implantation, diffusion coefficient of implanted atoms increases temporarily (i.e. until damage is annealed out) up to 10 times.
ion implantation damage, diffusion, diffusion coefficient

transistor three-terminal (1-2-3) semiconductor device in which input signal 1-2 (voltage or current depending on the type of transistor) controls output current 1-3; the most important semiconductor device; performs switching and amplifying functions; early on replaced bulky and inefficient vacuum triode in electronic circuits; invention of transistor (J. Bardeen, W. Brattain and W. Schockley, 1947) triggered electronic revolution after the World Word II; can operate as a discrete device or a building cell of an integrated circuit; numerous kinds of transistors can be distinguished based on their design and principles of operation; two major classes of transistors are unipolar (field-effect transistor, or FET) where Junction FET, Metal-Semiconductor FET and Metal-Oxide-Semiconductor FET are distinguished and bipolar transistors represented by Bipolar Junction Transistor, BJT.
bipolar transistor, unipolar transistor, Junction FET, Metal-Semiconductor FET, Metal-Oxide-Semiconductor FET, Bipolar Junction Transistor

transistor gain the ratio of the collector (terminal 3)-emitter (terminal 1) current (I_{CE}) and base (terminal 2)-emitter (terminal 1) (I_{BE}) current in the bipolar junction transistor; constant at the constant collector-emitter voltage; very small I_{BE} current controls much larger I_{CE} current.
transistor, bipolar transistor

transistor leakage term refers to undesired current that flows when transistor is turned off; in advanced MOS/CMOS devices using ultra-thin gate oxide leakage current is primarily a tunneling current across the gate oxide; transistor leakage must be minimize to prevent generation of excessive heat in logic ICs containing billions of transistors.
leakage current, tunneling, tunneling current, heat management

transistor scaling a process of scaling down transistor geometry for the purpose of its performance improvement in logic and memory applications; must be executed following certain rules to avoid deleterious short-channel effects.
scaling rules, transistor switching, gate length, short-channel effects

transistor switching a process of switching transistor from *"on"* to *"off"* state and vice versa; fundamental action in logic (computational) applications.
logic circuit

Transit-Time Diodes a group of semiconductor diodes designed for and broadly used in microwave applications.
BARITT, IMPATT, TRAPATT

Transmission Electron Microscopy, TEM a technique used to visualize and to allow quantitative evaluation of cross-sections of multi-layer nanostructures at the atomic level; requires tedious preparation of the very thin samples transparent to electrons, but offers unsurpassed capability in monitoring ultra-thin layers and interfaces; a powerful method allowing visualization of structural features at the nanometer level.
Scanning Transmission Electron Microscopy

transmission mask see *mask, transmission*.

transverse straggle standard deviation of the distribution of implanted ions in the direction normal to the direction of ions impinging on the surface of the wafer.
ion implantation, straggle

trap in semiconductor terminology this term commonly refers to the recombination site represented by the energy level in semiconductor bandgap in which minority carrier is trapped (captured) for a period of

time and then released (thermally "ejected"); this is in contrast to a recombination center in which prior to being released the minority carrier is annihilated by recombination with a captured majority carrier.
recombination, recombination site

trap assisted recombination a recombination process which is promoted by the traps (energy levels) in semiconductor's energy gap (bandgap); responsible for shortened minority carrier lifetime in defective semiconductors.
trap, recombination, minority carrier lifetime

TRAPATT diode Trapped Plasma Avalanche-Triggered Transit diode; a diode operating in a microwave range.
IMPATT diode

trench geometrical feature etched into semiconductor substrate; anisotropic etch in the direction normal to the surface; features high aspect ratio.
anisotropic etch, aspect ratio, trench isolation, trench capacitor

trench capacitor capacitor built into a trench etched in the semiconductor substrate; by using trench configuration area of capacitor can be expanded (capacitance increased) without increasing area of the chip needed to form a capacitor.
storage capacitor, DRAM

trench isolation a trench etched into the semiconductor substrate filled with oxide; such trench is located in between adjacent devices in an integrated circuit to provide electrical isolation between them.
shallow trench isolation

tri-gate transistor a 3D MOSFET in which vertical channel is contacted by the gate stack on its three sides.
pi gate, FinFET, gate all-around

trivalent silicon silicon atom in which one out of its four covalent bonds is unsaturated (broken); acts as a defect which can be electrically active.
dangling bond, interface trap

TSV see *Through Silicon Via.*

TTF Time-To-Failure.

TTL Transistor-Transistor Logic; type of logic gate implemented using bipolar technology.
Emitter Coupled Logic, Integrated Injection Logic

tungsten, W refractory metal (melting point above 2000 °C) used in silicon IC processing primarily to provide vertical connection (vias) between metal lines in multilevel interconnect scheme.
plug, via

tungsten silicide, WSi$_2$ contact (ohmic) material in Si technology; resistivity 30 - 70 $\mu\Omega$-cm; formed at the sintering temperature of 1000 °C.
silicide, ohmic contact, sintering temperature

tunnel (tunneling) barrier a potential barrier thin enough for the probability of electron to tunnel across it is very high and tunneling is a dominant transport mechanism; displays characteristics of the ohmic contact.
tunneling, sub-collector contact

tunnel diode two-terminal device in which current flow is controlled by the carriers tunneling across the potential barrier associated with *p-n* junction, or with ultra-thin oxide in MOS devices.
tunneling

Tunnel FET Tunnel Field-Effect Transistor; a MOSFET in which tunnel barrier is created at the source-channel contact in order to increase drive current of the transistor and subthreshold slope.
drive current, tunnel barrier, FET, subthreshold slope

Tunnel FinFET a Tunnel FET implemented in 3D FinFET configuration.
Tunnel FET, FinFET

tunnel junction see *tunnel diode*.

tunnel oxide an oxide in MOS structures so thin that the probability of electron direct tunneling across it (i.e. between metal gate and semiconductor) is very high; in the case of Si-SiO$_2$ based MOS devices significant direct tunneling current is observed for oxides thinner than

about 3.0 nm; for somewhat thicker oxides (~7 nm) Fowler-Nordheim tunneling may dominate.
direct tunneling, tunneling, Fowler-Nordheim Tunneling

tunneling, tunneling current   transport of electron across the potential barrier without changing its energy; as opposed to electron transport "over" the barrier (thermionic emission) in which case its energy must be changed; tunneling probability is a strong function of the width of potential barrier; dual nature of electron (particle and wave) comes to play in the quantum effect of tunneling; direct tunneling through the ultra-thin gate oxide is a factor limiting gate oxide scaling.
tunnel barrier, direct tunneling, Fowler-Nordheim tunneling, wave-particle duality

TUNNET   Tunnel Injection Transit-Time diode; GaAs based high-performance microwave transit-time diode.
IMPATT, TRAPPAT

turbomolecular pump   clean and efficient high-vacuum pump (below 10^{-7} torr) commonly used in state-of-the-art vacuum systems involved in semiconductor technology; very fast spinning turbine forces gas molecules out; can be used only after initial evacuation of air/gas using roughing pump is completed.
roughing pump

twin-well CMOS   CMOS structure using both n-type and p-type wells for better control of threshold voltage and transconductance of each transistor.
CMOS, well, transconductance, threshold voltage

two-dimensional electron gas (2-DEG)   a layer featuring very high accumulation of electrons confined in a very thin potential well within the heterojunction; electrons are free to move only in the other two directions, hence, "two-dimensional"; displays distinct electron transport characteristics allowing very fast transistors; e.g. HEMT.
quantum well, High Electron Mobility Transistor

two-dimensional (2D) material   see *quantum well, graphene, silicene.*
molybdenum disulfide

twin-gate transistor a MOSFET in which channel either planar or vertical is contacted by the gate stack on two sides.
double (dual) gate transistor; tri-gate transistor, planar CMOS, FinFET

TXRF see *Total Reflection X-Ray Fluorescence spectroscopy.*

TZDB see *Time-Zero Dielectric Breakdown.*

U

UHV Ultra-High Vacuum; typically vacuum below 10^{-8} torr.
torr

UHV CVD Ultra-High Vacuum Chemical Vapor Deposition; carried out at the pressure $\sim 10^{-8}$-10^{-9} torr; used to deposit epitaxial layers (e.g. SiGe) at low-temperature (around 500 °C).
CVD, epitaxial deposition

UJT unijunction transistor.

ULPA filter Ultra-Low Penetration Air filter; used in high-end cleanrooms; removes 99.999% of all particles 0.1 - 0.2 μm and larger from circulated air; higher performance than HEPA filters.
HEPA filter, cleanroom

ULSI Ultra Large Scale Integration; integrated circuits with over million transistors per chip and minimum feature size typically below 100 nm.
VLSI

ultra-pure water see *deionized water.*

ultra-shallow junction term refers to the *p-n* junction forming source and drain in the MOSFET; typically less than 40 nm deep, but can be as shallow as < 10 nm; MOSFET' scaling rules require ultra-shallow junctions to limit the short-channel effects.
short-channel effects, MOSFET, raised source-drain

Ultra-Thin Body SOI, UTB SOI the SOI substrate with Si active layer thinner than 10 nm; needed to implement fully depleted CMOS on SOI substrates; same as ET (Extremely Thin) SOI.
Extremely Thin SOI, fully-depleted SOI

ultra-thin oxide typically thinner than 3 nm; the gate oxide/dielectric in most advanced CMOS based ICs; direct tunneling current between metal gate and semiconductor substrate through ultra-thin oxide adversely affects performance of MOS devices.
gate dielectric, tunnel oxide, tunneling

Ultra-Thin Silicon on Sapphire, UTSOS ultra-thin film of Si is epitaxially grown on sapphire; Silicon-On-Sapphire, or SOS, is a type of the SOI substrate; due to the lattice mismatch deposited Si is in the pseudomorphic state and is highly strained; if formed on such substrates the CMOS based circuitry would feature superior performance.
silicon on sapphire, strained lattice, pseudomorphic material

ultrasonic agitation, ultrasonic scrubbing used to enhance cleaning operations or to improve homogeneity of liquid solution, e.g. uniformity of dispersion of nanoparticles in colloidal solutions; sonic energy at the frequency range of 10 - 100 kHz is applied to the liquid in which wafers are immersed; not as effective in commercial wafer cleaning processes as megasonic agitation, and hence, not used.
megasonic agitation

ultraviolet, UV radiation in the electromagnetic spectrum featuring wavelength λ in the range from about 150 nm to about 400 nm (corresponding energy in the range from about 8.2 eV to about 3.1 eV); carries enough energy to stimulate photochemical reactions in solids (e.g. photoresist exposure) and/or in gases (e.g. Photo-CVD) in semiconductor processing.
Photo-CVD, photolithography

Ultraviolet (UV) Electron Spectroscopy, UPS a method of material characterization; based on the emission of photoelectrons from the solid stimulated by UV irradiation; an ESCA method (along with XPS).
ESCA, XPS

U-MOSFET, UMOS power MOSFET with channel built around U-shaped trench etched into semiconductor substrate; extension of the V-MOSFET concept; features lower than other power MOSFETs (D-MOSFET, V-MOSFET) "on" resistance.
power device, V-MOSFET, D-MOSFET

undercut undesired result of the isotropic etching process; etched pattern extends lateraly under the photoresist limiting accuracy of the pattern transfer process.
isotropic etch, etching

uniaxial strain an uniaxial strain in the crystal has stress acting in one direction; uniaxial strain created in the channel of the NMOSFET has beneficial effect on electron mobility in the channel, and hence, transistor performance; engineering of strain in the channel is an important part of the state-of-the-art silicon CMOS technology.
biaxial strain, tensile strain

unipolar device semiconductor device which operation is based predominantly on majority charge carriers; e.g. all transistors based on the field effect fall into this category; alternative to bipolar device.
Schottky diode, Field-Effect Transistor

unipolar transistor synonymous with Field Effect Transistor, FET, i.e. transistor which operation is controlled by the majority carriers (majority carrier current in the channel).
Field-Effect Transistor, bipolar transistor

unit cell the smallest building block of crystal lattice; when reproduced three dimensionally forms a crystal.
crystal, crystal lattice

unlimited source diffusion see *pre-deposition.*
limited source diffusion

unsaturated bond occurs when atom in a solid is missing and as a result the bond of the adjacent atom is not neutralized; in a single-crystal solid such situation is synonymous with a point defect; on the surface, unsaturated bond needs to be rendered passive (surface passivation) in order to stabilize its chemical reactivity.
dangling bond, point defect, trivalent silicon, surface passivation

UPS see *Ultraviolet Electron Spectroscopy.*

USJ see *Ultra-Shallow Junction.*

UTB SOI see *Ultra-Thin Body SOI.*

UV see *ultraviolet.*

UV/chlorine (Cl₂) UV irradiation of Si surface in Cl_2 ambient at the reduced pressure and wafer temperature not exceeding 200 °C; part of the silicon gas-phase (dry) cleaning scheme; causes volatilization of most of the metallic contaminants on Si surface and slight etching of silicon.
dry cleaning, metallic contaminant

UV cleaning gas-phase (dry) cleaning process stimulated by UV radiation; e.g. volatilization of organic contaminants by UV/ozone.
cleaning, UV/ozone, ozone

UV, deep see *deep UV.*

UV, extreme see *extreme UV.*

UV/ozone cleaning dry (gas-phase) cleaning processes used to remove organic contaminants from solid surfaces by oxidation; UV irradiation in the ambient containing oxygen generates ozone which is a very strong oxidizing agent; spectrum of UV radiation should contain 185 nm and 254 nm wavelengths.
organic contaminant, ozone

V

vacancy a point defect in crystalline solids; a missing atom (empty site) in single crystal lattice; also know as Schottky defect.
crystal defect, point defect

vacuum a space entirely empty of matter; in practice a space in which the pressure is by orders of magnitude lower than atmospheric pressure; (atmospheric pressure = 760 torr, high-vacuum 10^{-6} - 10^{-8} torr); broadly exploited medium in semiconductor manufacturing.
torr

vacuum gauge an instrument used to measure pressure below atmospheric; key component of vacuum based semiconductor processing tools; different types of gauges need to be used in different vacuum ranges.

vacuum level defines threshold energy at which electron can leave an atom; a reference used in the definition of the work function.
work function

vacuum pump an instrument used to reduce pressure below atmospheric pressure (to create vacuum); different pumps are used for different vacuum ranges; performance defining component of semiconductor vacuum tools.
turbomolecular pump, cryogenic pump, diffusion pump, mechanical pump, roughing pump

vacuum ultraviolet, VUV an ultra-short wavelength UV radiation emitted by plasma; e.g. 13.5 nm wavelength used in Extreme UV Photolithography.
ultraviolet, extreme UV, Extreme UV photolithography

valence band energy band in semiconductor that is completely filled with electrons at 0 K; lack of empty states inhibits any motion of electrons, hence, electrons cannot conduct current in valence band; to contribute to electrical conductivity electrons must acquire energy sufficient to move from the valence band to the conduction band; in semiconductors and insulators valence band is separated from the conduction band by the energy gap (bandgap); in metals valence band and conduction band overlap.
conduction band, energy gap

Van der Pauw method alternative to conventional four-point probe method used to measure resistivity of semiconductors; advantages: compatible with thin-films and oddly shaped samples.
four-point probe, resistivity

Van der Waals forces (interactions) a broad term describing common in nature attractive forces between molecules or groups of atoms; do not involve electrons sharing and forming a strong bond (e.g. covalent bond); as an example, graphene layers are connected by Van der Waals forces to form graphite.
covalent bond, graphene

Van der Waals heterostructure formation of heterostructures comprised of different 2D materials (e.g. graphene and molybdenum disulfide) kept together by Van der Waals forces.
heterostucture, graphene, molybdenum disulfide, Van der Waals forces

varactor "variable reactor"; variable reactance semiconductor device; *p-n* junction diode designed to take advantage of the variations of its reactance; used as an active element in parametric circuits.

variable shape beam scanning mode in e-beam lithography in which shape of the beam is changing depending on the geometry of exposed area.
electron-beam lithography, fixed shape beam, raster scan, vector scan

varistor "variable resistor"; semiconductor devices featuring non-linear current-voltage characteristics; basically any semiconductor diode is a varistor.
diode

VCSEL Vertical Cavity Surface Emitting Laser; in short Surface Emitting Laser (LSE); semiconductor laser in which generated radiation is propagated vertically, i.e. in the direction normal to the semiconductor surface and *p-n* junction plane rather than in the direction parallel to the surface and *p-n* junction plane (conventional operation mode).
laser, edge emitting laser, surface emitting laser

VDMOSFET Vertical Double-Diffused MOSFET; power transistor structure; variation of D-MOSFET.
D-MOSFET

vector scan scanning mode used in electron-beam lithography in which beam is scanning selected areas only; after scanning of selected area is completed beam is turned off and moved to another area to be scanned; more efficient scanning mode as compared to raster scan.
electron-beam lithography, raster scan, variable shape beam

velocity overshoot occurs when the average velocity of carriers in semiconductor exceeds saturation velocity; possible during ballistic transport.
ballistic transport, velocity saturation

velocity saturation see *saturation velocity.*

vertical channel channel in the MOSFETs that is not in the plane of the substrate wafer; it can be a vertical channel formed on the surface of the substrate wafer in logic IC transistors (e.g. FinFET) or vertical channel etched into the substrate wafer in power transistors (e.g. U-MOS FET). *FinFET, U-MOSFET, V-MOSFEF*

vertical diffusion diffusion in solids in which diffusing species are moving in a direction perpendicular to the solid (wafer) surface. *diffusion, lateral diffusion*

vertical furnace a furnace in which process tube is located vertically, i.e. boat with wafers moves up and down; used in semiconductor batch processes for oxidation, CVD, diffusion, and anneals; batch process; superior to horizontal furnace in terms of the smaller foot print in the clean room, compatibility with automated wafer loading, and heating uniformity. *furnace, horizontal furnace, batch process*

Vertical-Slit Field Effect Transistor, VeSFET a vertical transistor architecture which efficiently integrates transistor with multi-level metallization scheme in high-density ICs; superior flexibility in logic/memory circuit design; junctionless with two independently controlled gates and featuring advantegous I_{on}/I_{off} characteristics; offers alternative to transistor scaling solution to the challenges of ultra-low power digital ICs without a need to generate patterns below 65 nm. *digital ICs*

vertical transistor bipolar transistor in which current across emitter-base and collector-base junctions flows in direction normal to the wafer surface; conventional BJT configuration as opposed to lateral transistor which features inferior characteristics. *lateral transistor, bipolar junction transistor*

vertical tunnel FET tunnel FET in vertical configuration. *tunnel FET*

VeSFET see *Vertical-Slit Field Effect Transistor.*

VeSTICs integrated circuit based on the Vertical Slit Field Effect Transistors.

V_{FB} **(flat-band voltage) roll-off** see *roll-off.*

VHSIC Very High-Speed Integrated Circuit.

via high aspect ratio opening etched in an interlayer dielectric which is then filled with metal, usually tungsten, to provide vertical connection between stacked up interconnect metal lines; also hole etched through the thinned Si wafer in TSV technology.
inter-layer dielectric, plug, Through Silicon Via, three-dimensional integration, aspect ratio

via first term refers to the 3D packaging in which through silicon via (TSV) is implemented after wafer thinning, but before chips bonding.
three-dimensional integration, through silicon via, via last

via last term refers to the 3D packaging in which TSV is implemented after chips thinning and bonding.
three-dimensional integration, through silicon via, via first

via veil "veil"-like residue resulting from the resist stripping process following via etch in multilevel interconnect scheme.
resist stripping, via

VLS growth Vapor-Liquid-Solid growth; can be used for instance to form Si nanowires on Si substrate.
nanowire

VLSI Very Large Scale Integration; integrated circuits with 100,000 to 1 million transistors/chip and minimum feature size between approx. 1.5 μm and 0.6 μm.
ULSI

V-MOSFEF, VMOS a discrete vertical-channel power MOSFET structures; in V-MOSFET channel is built around V-shaped groove etched (employing preferential etching technique) in silicon substrate; alternative technologies: DMOSFET and U-MOSFET.
preferential etching, U-MOSFET

volatile memory a memory in which stored information is lost when power to the memory cell is turned off.
non-volatile memory

volume defect voids and/or local regions featuring different phase (e.g. precipitates or amorphous phase) in crystalline materials.
crystal defects

volume properties see *bulk properties*.

VPD Vapor-Phase Decomposition.

VPD-AAS Vapor-Phase Decomposition Atomic Absorption Spectroscopy; method used to determine concentration of inorganic elements, mainly metals on the wafer surface.

VUV see *Vacuum Ultraviolet*.

W

wafer in semiconductor terminology term "wafer" has various meanings: *(i)* thin (thickness depends on wafer diameter, but is typically less than 1 mm), circular slice of homogenous single-crystal semiconductor material used to manufacture semiconductor discrete devices, electronic and photonic, and integrated circuits; depending on material, wafer diameter may range from about 25 mm to 450 mm; *(ii)* SOI wafer, i.e. silicon wafer as above with very thin SiO_2 layer buried at certain distance from the wafer surface, *(iii)* very thin (200 μm and thinner) square (typically ~150 mm on the side) slice of multicrystalline silicon used in solar cell fabrication, *(iv)* non-semiconductors wafers used as substrates in semiconductor device engineering, e.g. sapphire, quartz, or glass, circular or square depending on application.
bulk wafer, SOI wafer, multicrystalline material, silicon-on-sapphire

wafer bonding process in which two semiconductor wafers are bonded to form a single substrate featuring desired characteristics; commonly applied to form SOI substrates; bonding of wafers of different materials (different chemical composition and/or crystallographic structure), e.g. GaN on Si, or SiC on Si allows development of novel device structures; also applies to temporary bonding of semiconductor wafers with

semiconductor and non-semiconductor substrate wafers where one wafer is used as a handle wafer.
bonded SOI, SOI, wafer de-bonding, handle wafer

wafer charging process of acquiring static electric charge by the wafer during processing; effect that needs to be monitored and minimized.
static charge

wafer damage term refers to the mechanical damage to the semiconductor wafer incurred during processing; e.g. chipping during wafer handling operations.

wafer de-bonding temporarily bonded wafers can be de-bonded, i.e. separated, without being damaged, using various techniques which depend on the bonding method.
wafer, wafer bonding

wafer diameter term refers to the diameter of the single-crystal semiconductor wafers used to manufacture electronic and photonic devices/ circuits; varies from ~25 mm for some difficult to obtain in the single-crystal form compound semiconductors to 450 mm for silicon.
single-crystal growth, wafer fabrication

wafer fabrication process in which single-crystal semiconductor in the form of a rod (ingot) or a boule or a multicrystalline semiconductor slab is transformed by cutting (wafering), grinding, polishing, and cleaning into a circular (with desired diameter) or square wafers used to manufacture semiconductor devices.
wafer, wafering, wafer diameter, double-side polishing

wafer flat a flat area machined at the edge of the single-crystal semiconductor wafer; location and number of wafer flats contains information on crystal orientation of the wafer and dopant type (*n*- or *p*-type).
wafer

wafer mark in high-end semiconductor manufacturing each wafer is identified with a laser scribed bar code; in this way each wafer can be identified and monitored throughout the entire manufacturing process.
wafer

Wafer Scale Integration, WSI an integrated circuit is not limited to a single chip, but is spread over the entire wafer; in effect, the whole wafer acts as a chip.
chip

wafer thinning reduction of the wafer thickness typically by hundreds of μmeters; employed e.g. in SOI fabrication by wafer bonding and in 3D integration
Chemical-Mechanical Polishing, three-dimensional integration, Through Silicon Via

wafer track term typically refers to a track-like set up which integrates several tools needed to processes photoresist (deposition, soft bake, exposure, developing, hard bake) in advanced photolithographic processes.
photoresist, photolithography

wafer warpage highly undesired deformation of the processed wafer due to the stress that can be introduced by aggressive thermal treatments; sometime referred to as "potato chip effect"; may interfere with photolithographic processes by affecting depth of focus in various parts of the wafer.
Wafer, depth of focus

wafering process of cutting single-crystal ingots (rods) or boules or slabs of semiconductor materials into wafers used in semiconductor device manufacture; synonymous with "slicing".
wafer, slicing

water marks readily identifiable marks left on the surface of the wafer as a result of ineffective wafer drying process; likely to promote local agglomeration of particles on the surface that may cause catastrophic damage to the device/circuit being processed.
drying, particle

wave-particle duality the principles of quantum physics supported by experimental evidence show that light behaves not only as wave, but also as a particle while particle (e.g. electron) behaves not only as particle, but also as a wave; electron's behavior as a wave is evidenced in various physical phenomena observed in highly geometrically confined semi-

conductor structures in the presence of high electric field, e.g. ballistic transport and tunneling.
electron, ballistic transport, tunneling

waveguide in semiconductor engineering an optical waveguide is a thin-film structure designed to guide light (photons) between elements of the circuitry; in the same way plasmonic waveguides are designed to guide plasmons.
photon, plasmon

wearable semiconductors terms refers to semiconductor electronic and photonic devices/circuits performing communication, navigation, sensing, monitoring etc. functions that can be integrated with clothing, worn as self-contained devices or permanently attached to the skin.
flexible electronics and photonics

Weibull plot convenient way of representing statistics of failure events in semiconductor devices and materials; e.g. distribution of oxide break-down events as a function of electric field; cumulative failure F is plotted as $-\ln(1\text{-}F)$ vs. measurement variable such as electric field.
breakdown

well term commonly refers to implanted/diffused region in Si wafer needed to implement complementary MOS (CMOS) cell; depending on the design n-well, p-well or both n-well and p-well (twin well) are implemented.
twin-well CMOS

wet bench typically fully automatic process tool used to carry out wet cleaning and etching operations in semiconductor processing; commonly includes several tanks each containing either cleaning/etching solution or deionized rinsing water in which wafers are immersed in predetermined sequence; commonly equipped with megasonic system; also includes drying module.
cleaning, drying, immersion cleaning, megasonic agitation

wet cleaning process of contaminants removal from the wafer surface in the liquid-phase; prevailing cleaning method in semiconductor manufacturing; wet cleaning chemistries are selected to form soluble compounds of surface contaminants or to promote dislodging of the contaminant (e.g. particle); often enhanced by megasonic agitation;

always followed by deionized water rinse and dry cycle; implemented in the variety of ways using variety of cleaning chemistries; batch or single-wafer process.
APM, HPM. SPM, dry cleaning, immersion cleaning, spray cleaning, spin cleaning, megasonic agitation

wet etching etching process in semiconductor processing relying on chemical reaction in the liquid phase; by definition highly isotropic; can be also very selective; under certain conditions wet etching can be much faster along selected crystallographic planes.
etching, dry etching, isotropic etching, selective etching, preferential etching

wet oxidation thermal oxidation of silicon carried out in water vapor containing oxygen, or in steam; significantly higher oxide growth rate as compared to dry oxidation; used to form relatively thick films of SiO_2 on Si surface; wet oxide contains Si-H and Si-OH, and hence, features inferior to dry oxide electrical integrity.
dry oxidation, thermal oxidation

wetting angle same as contact angle; tangent angle at the interface between droplet of liquid and a solid surface; measure of the surface energy; 0° for hydrophilic surface and 90° for hydrophobic surface; contact angle is very sensitive to the condition of the surface; very effective in monitoring variations in chemical and physical characteristics of the surface.
hydrophylic, hydrophobic

WF see *work function.*

white LED there are no LEDs generating white light; white light LEDs are constructed either by using several single-color LEDs to produce wavelength mix that appears white or by using a short-wavelength (e.g. blue) LED in conjunction with phosphors.
light emitting diode, phosphor, lighting

wide-bandgap semiconductor semiconductor featuring energy gap (bandgap) E_g wider than about 2.5 eV; useful in high temperature applications and emission of blue light; in this last case bandgap needs to be direct; e.g. SiC (E_g = 2.9 eV, indirect bandgap), GaN (E_g = 3.5 eV, direct bandgap), ZnS (E_g = 3.68 eV, direct bandgap).
gallium nitride, silicon carbide, zinc sulfide

wire bonding part of the chip assembly process in which very thin (about 30 µm in diameter) gold wire is used to connect chip and the package; bonding pads at the perimeter of the chip are connected to the lead connections in the package; typically uses combination of thermal compression and ultrasonic motion.
assembly, package

work function energy needed to move electron in the atom from Fermi level to vacuum level, i.e. to outside of the atom; commonly used as a reference while comparing energy state of various materials and predicting electrical properties of the contact between them.
Fermi level, vacuum level

work function difference difference in work function between two materials in physical contact; defines characteristics of contact between two materials featuring different work functions; in conductor-semiconductor contact work function difference determines height of the potential barrier, and hence, determines whether contact is ohmic or rectifying; in MOS devices work function difference between semi-conductor and gate contact affects threshold voltage; should be as low as possible.
ohmic contact, potential barrier, rectifying contact, work function

wrapped around gate see *gate all-around.*

WSI see *Wafer Scale Integration.*

WSP Wafer Stack Package; type of package in 3D chip stacking technology.
three-dimensional integration

wurtzite lattice hexagonal close-packed crystallographic structure; while most semiconductors crystallize following cubic structure, some, most notably nitrides (e.g. AlN, BN, GaN) as well as certain polytypes of SiC, feature wurtzite lattice.
cubic system, zincblend lattice, hexagonal crystal system, nitrides

X

XPS see *X-Ray Photoelectron Spectroscopy.*

X-ray diffraction, XRD material characterization method used to determine crystallographic structure of solids based on the diffraction of X-rays irradiating the solid; detects crystal defects and alien phases in the solid; not sensitive to chemical composition of analyzed solid.

X-ray Fluoroescence, XRF a method used to study chemical composition of solids; does not distinguishes between surface and the bulk of the solid, and hence, is not useful in characterization of semiconductor surfaces; total reflection version of XRF (TXRF) was devised to fulfill this last function.
Total Reflection X-Ray Fluorescence spectroscopy

X-ray lithography, XRL lithography method which uses X-rays to exposed the resist; due to shorter wavelength of X-ray radiation (0.4 - 4 nm), XRL allows higher resolution of the pattern transfer than photolithography which uses longer wavelength UV irradiation; XRL requires special masks and resists sensitive to X-rays; not used in mast production due to the complexity and cost of high-intensity X-rays sources needed in lithographic applications as well as X-ray masks.
lithography, synchrotron radiation, synchrotron, X-ray masks

X-ray mask see *mask, X-ray lithography*.

X-Ray Photoelectron Spectroscopy, XPS surface analysis method used to determine chemical composition of solid surfaces; allows determination of bonding energy; analysis is based on the determination of energy of electrons emitted from the solid as a result of irradiation with monochromatic X-rays; very useful in detecting on Si surface Si-suboxides, as well as O, F, C; an ESCA method (along with UPS).
Electron Spectroscopy for Chemical Analysis, Ultraviolet Photoelectron Spectroscopy, suboxide

X-ray resist resist used in X-ray lithography; X-ray resist.

XSC see *excitonic solar cells*.

Y

yield in semiconductor terminology synonymous with "manufacturing yield", i.e. number defining percentage of fully operational devices (chips) out of all devices (chips) manufactured; must be sufficiently high to warrant profitability of the manufacturing process.

yield ramping increase of manufacturing yield with time, i.e. as a given process becomes more mature.

Z

Zener breakdown reverse-bias breakdown of the highly doped *p-n* junction triggered by Zener effect.
breakdown, avalanche breakdown

Zener diode a *p-n* junction based diode exploiting Zener effect.
Zener effect

Zener effect occurs in heavily doped *p-n* junctions at high reverse-bias voltage; energy bands in a junction region are sloped in such way that electron may tunnel directly from the valence band in *n*-type region to the conduction band in *p*-type region causing rapid increase of current; may cause breakdown of *p-n* junction but also can be used in a controlled fashion to make a functional device (Zener diode).
p-n junction, direct tunneling

Zener tunneling see *Zener effect*.
tunneling

Zerbst plot a way of plotting capacitance *vs.* time relationship for MOS capacitor pulsed into deep depletion; allows determination of generation and recombination lifetime in semiconductor substrate.
deep depletion, generation lifetime, recombination lifetime

zero-dimensional (0D) material see *nanodot, nanocrystalline quantum dot, one-dimensional material, two-dimensional material, three-dimensional material*

zeta potential potential of solid surface interacting with an ambient featuring specific chemical composition; plays important role in surface cleaning and conditioning processes as it defines the way semiconductor surface interacts with any given chemistry.
surface conditioning, wet cleaning

zigzag GNR a type of graphene nanoribbon (GNR) which can display metallic properties; term "zigzag" refers to the specific configuration of carbon atoms at the edge of the nanoribbon.
graphene, armchair GNR

zinc-blend lattice crystal structure which belongs to the face-center cubic (FCC) crystal family; most of the III-V compound semiconductors feature zincblend lattice.
cubic system, diamond lattice, face-centered cubic, wurtzite structure

zinc oxide, ZnO wide-bandgap II-IV group semiconductor; $Eg = 3.4$ eV; direct; possible blue and violet emission and UV detection; features high transparency to white light, hence, allows transparent electrodes; due to the wurtzite lattice structure and similar lattice constant to GaN can be used as a substrate for GaN (lattice matched to InGaN with 22% In content); successfully used in TFT technology.
direct bandgap semiconductor, wide-bandgap semiconductor, thin-film transistor

zinc selenide, ZnSe II-VI group semiconductor; bandgap $E_g = 2.7$ eV, direct; zinc blend or wurtzite crystal structure; can be used for blue light emission and UV detection.

zinc sulfide, ZnS wide-bandgap II-VI group semiconductor; zinc blend crystal structure; the largest bandgap ($E_g = 3.68$ eV) among semiconductors considered for practical applications (only diamond and boron nitride feature wider bandgap); direct bandgap makes ZnS potentially attractive as a blue light emitter and UV detector.
direct bandgap semiconductor, wide-bandgap semiconductor, diamond, boron nitride

zinc telluride, ZnTe II-VI group semiconductor; bandgap $E_g = 2.26$ eV, direct; typically zinc blend crystal structure but possible also in wurtzite structure; can be used for blue light emission as well as in solar cells.
solar cell

zirconium oxide, zirconia, ZrO₂ dielectric featuring dielectric constant k in the range of 22-25; tends to crystallize at about 700 °C; considered as an alternative to SiO_2 gate dielectric in CMOS technology; in many aspects similar to hafnium oxide.
gate oxide capacitance, hafnium oxide, high-k dielectrics

zirconium silicate, ZrSiO₄ dielectric material featuring dielectric constant $k \sim 16$; in contact with Si combines characteristics of silicon dioxide, SiO_2, and zirconium oxide, ZrO_2; considered as an alternative to SiO_2 gate dielectric in CMOS technology; in many aspects similar to hafnium silicate.
gate oxide capacitance, hafnium silicate, high-k dielectric

zone purification see *zone refining*.

zone refining a method allowing purification of crystalline solids; based on the same principle as float-zone crystal growth; results in accumulation of impurities at one end of the processed solid rod.
float-zone crystal growth

Z-RAM™ Zero-capacitor Random Access Memory; floating body dynamic memory technology which unlike DRAM does not require storage capacitor; both switching and storage functions are carried out by a single transistor formed on SOI substrate.
DRAM, RAM, SOI

Printed in the United States
By Bookmasters